Back to Basics

Aircraft Handling, Cockpit Mechanics, and
Flight Procedures

By the Editors of *Flying* Magazine

Idiom Press
Box 583
Deerfield, IL 60015

To the student pilot-
with ten hours or ten thousand

Publisher's Note: This book was originally published in 1977, and became an immediate classic. No pilot, then or now, could read this book without learning a few things − or a lot. And not just by rote, either. The clear, lucid explanations lead the reader into a real and expanded understanding of the subject matter.

Unfortunately, shortly after the book was published, the original publisher, a major New York book publisher, withdrew their interest and participation in the field of aviation. *Back to Basics* went out of print and has since been unavailable, until *Flying* Magazine graciously licensed publication rights to Idiom Press.

Can an aviation book originally published in 1977 be valid today? The answer is a resounding yes! Of the 43 chapters in this book, one, dealing with Federal Airspace structures, is obsolete. Two chapters are obsolescent, lacking recent developments, but still offer valuable and thought provoking considerations of the material covered. The balance are as timely today as when they were first written.

Pilots who read this book will be surer and safer, and will better understand how to operate their planes in a safe, efficient and economical manner.

Idiom Press
Box 583
Deerfield, IL 60015

CONTENTS

FOREWORD

Whatever their licenses say, all plots are student pilots. As long as we live and we never stop learning — learning things we didn't know and relearning what we have forgotten.

The *Back to Basics* series was aimed at students at every level of experience. We hoped to make the articles instructive to the novice while keeping them interesting to the seasoned pro. To our mild surprise, the series, despite its low-key approach to a field that tends to be bloated with mystique, was enthusiastically received. Students wrote to thank us for making clear what had been incomprehensible; instructors congratulated us on being the first to present simple, precise, and, above all, *correct* explanations of concepts such as laminar flow and manifold pressure (which are so basic that they normally get overlooked entirely and remain forever an abyss of mystery within a pilot's knowledge); oldtimers wrote to tell us we were crazy, downwind turns *were* different. We tried to be thorough, systematic, exhaustive, scientifically correct — and even interesting. For the most part, we think we succeeded, and that is why we have gathered the series (with some supplementary articles of similar format to fill out the occasional blanks) into a book. The knowledge that it contains is the best that a group of specialists in flying and in writing about flying have had to offer; it may amuse, it may instruct, it may occasionally save lives. We don't want it to get lost.

Robert B. Parke
Publisher, *Flying* Magazine

I. THE AIRCRAFT

It took mankind several hundred years of trying — and several thousand of getting ready to try — to find out how to build a machine that could fly. Today, any child knows how to make a sheet of paper into a creditable imitation of a bird.

People who ride in airplanes are usually indifferent to their mysteries. Pilots cannot afford to be so cavalier: in the air, ignorance is not always bliss. If a piece balks or breaks, it had better not be the piece that passeth understanding.

The first five articles in this section concern the physics of flight and how some of the parts of an airplane work. They offer knowledge more or less for its own sake. The remaining articles are quite practical: they are about the stuff of which one's daily dealings with airplanes are made.

1.

THE AIRPLANE,

OUTSIDE-IN

When you've been living day and night with those great frauds, automobiles, it's a pleasure to take a look at an airplane. Airplanes unite craft and feeling in a really beautiful way, rather like chess: as in chess, there are only so many moves that you can make, and everything that unfolds on the board, however strange, is accountable to a small set of rules. Everything that unfolds on an aeronautical engineer's drawing board is also accountable to rules — and to make the game more challenging, some of the rules are only partially known. What tells you whether or not you have complied with the rules is the performance of the airplane, and every airplane therefore contains the imprint of the designer's struggle to attune his intuition to the vagaries of the all-important, invisible element: air. The air is always there, making its obstinate demands, and there are also the demands of practicality, of the human cargo, of styling and marketability, cost and serviceability. Every airplane, designers say, is a group of compromises flying in tight formation. The beautiful artistry of the plane lies in the way in which the designers have eluded the traps set for them by cramping circumstance in order to follow, sometimes with perfect agility, the difficult calculus of the air so that, rather than appearing as a set of compromises, the airplane seems to radiate a profound singleness of intent.

Small airplanes for some reason are often rather gaudily painted — it seems to be an unquestioned practice in the industry — and their shapes are broken up as if by camouflage; they are also made, in some cases, to look trivial and toylike. The appearance is deceiving: they are in fact cunningly contrived, utterly serious, and built with an attention to detail so scrupulous that the phrase "aircraft quality" means that each kind of material, piece of hardware, and manufacturing practice used is the best, most carefully chosen, and most expensive available. Beneath that toylike, striped exterior beats a heart of chromium-molybdenum steel, normalized, heat-treated to 35 Rockwell hardness, magnafluxed, and cadmium-plated.

Almost all airplanes built today have the same sort of basic structure: a skin of thin aluminum sheet draped over frames, spars, ribs, and stiffeners also made of aluminum, the whole being riveted and bolted together. This type of construction was first adopted by the industry in the 1930s and has prevailed ever since. Only very recently have a significant number of airplanes begun to deviate from this traditional pattern: ultra-high-speed jets made of titanium or stainless-steel honeycomb to resist the heat generated by flight at several times the speed of sound and, at the extreme, sailplanes made of fiberglass laminates, capitalizing on the smoothness and freedom of form that plastics permits to attain the ultimate in streamlining. The peculiarity of stiffened-sheetmetal construction — or *semi-monocoque*, as it is also called — is that, because flight requires minimum weight, the individual parts of the metal structure are incredibly thin. Skins on some control surfaces may be as little as 16/1000" thick, or about as thick as five of these pages pressed together. Such sheetmetal is very flimsy to the touch, but if it is curved, its unsupported surfaces punctuated with stiffeners, and its free edges bent or riveted down, it is transformed in a way that seems astonishing even after you have seen it many times into hollow shells of tremendous rigidity as well as lightness.

Thin sheetmetal can make a light, rigid, and streamlined shell, but it is unsuited to carrying the powerful bending loads that occur in the wings and tall surfaces, which are relatively long, shallow beams supported along their entire length by the air and which in turn support at their centers the concentrated load of the fuselage and engines. The wings and tail therefore contain strong spars that resemble the I-beams used in buildings and bridges; these beefy aluminum spars resist strong bending loads, while the thin skin of the wing resists twisting. Interrupt this hollow shell with a few Plexiglas windows (glass is too heavy), garnish its more sensuously curved extremities with fiberglass fairings (it is sometimes ruinously costly to form aluminum into suitably streamlined shapes), and you have the essentials of an airplane.

So far, all you have provided for is structural strength and aerodynamic properties; you still need a powerplant, control systems, landing gear, instruments, and means of communication. Throughout most of the history of aviation, designers have taken the path of least resistance and placed airplane engines in the nose of the fuselage or along the front edge of the wing. Engines can also be put elsewhere — amidships, for instance, or on the rear end of the fuselage, driving a propeller that pushes rather than pulls the airplane — but for many reasons the front end is a convenient compromise. The engines of all light airplanes today are of the flat, opposed, air-cooled type, which means that, like the engine of a Corvair or Volkswagen, the cylinders are arranged on opposite sides of the engine to drive a propeller. The cylinders are provided with cooling fins like those of a motorcycle engine. Air is usually admitted to the engine compartment through holes in the front of the cowling, routed over the top of the engine and down among the cylinders, and expelled through vents on the underside of the cowling.

Propellers, usually two-bladed, are forged and hand-shaped from aluminum and are about 7' in diameter; three-bladed props are becoming popular, however, because they are slightly quieter and because they evoke the image of larger, more powerful airplanes that use multibladed props by necessity.

The engine gets its fuel, which it may use at a rate of from 5 to 20 gallons per hour (typical figures for 100- and 400-horse power engines, respectively, which represent the extremes of piston-engine sizes now in use), from tanks that are usually located in the wings, either in bomb-shaped tip tanks, in tanks within the wing structure, or in both. One of the many advantages of a all-metal construction is that parts of metal wings may be sealed off and used as fuel tanks, eliminating the need for an additional fuel-containing cell; still, many airplanes do use a rubber bag within the wing shell to contain the fuel, accepting a weight penalty in exchange for convenience of manufacture. Typically, a light airplane carries enough fuel for a least 4 hours of flight at cruising power; some can remain aloft for as long as 10 hours.

Most airplanes are equipped with three sets of movable control surfaces with which the pilot directs the course of the plane through the air. These are the ailerons, which are located on the outer portion of the rear edge of each wing; the elevators, which are the rear portions of the horizontal tail surfaces (though many airplanes have fully movable horizontal tails, called *stabilators*); and the rudder, which is the rear portion of the vertical tail surface. Each of these surfaces causes its part of the airplane to move away from the direction in which the control surface is deflected: if the elevators are raised, the tail of the plane is pushed downward; if the rudder swings to the right, the tail swings to the left; if the right aileron goes down while the left one comes up (they always work simultaneously but in opposite directions), the airplane rolls to the left.

The rudder is controlled by foot pedals, on which the landing gear wheel brakes are usually also mounted; the ailerons and elevators are controlled by the wheel, or yoke, which resembles the steering wheel of a car. Pushing the wheel forward raises the tail and makes the airplane dive; pulling it back lowers the tail and makes the airplane climb; turning the wheel to the right or left makes the airplane roll in the corresponding direction: like a bicycle, the airplane turns in the direction in which it tilts. The rudder is used primarily as a "trimming" control to keep the back of the airplane lined up behind the front, since in certain maneuvers the tail tends to swing out slightly to one side or the other. The ailerons, however, not the rudder, make the airplane turn.

In addition to these control surfaces there is usually one other pair of movable surfaces, called *flaps*, which are located on the inboard portion of the rear edge of the wing — the part not occupied by the ailerons. The flaps are hinged to deflect downward; together they increase the lifting ability of the wing as well as its drag, making possible better control of the glidepath during the landing approach and lower speed for the landing itself.

The control surfaces and flaps are usually operated by steel cables, though in some airplanes hollow aluminum pushrods are used instead. Cables are widely preferred because they are easily routed through the structure, but pushrods have a more positive feel, like the rack-and-pinion steering of a sports car, and they never need adjustment.

The prevalent type of landing gear on modern light airplanes is the "tricycle," which has one wheel in the front and two in the vicinity of the wings. In lower-cost airplanes the landing gear is fixed and simply protrudes from the airplane in flight; it produces quite a bit of drag, however; and a 200-mph airplane might lose 30 mph or more due to a fixed landing gear. In more expensive, higher-performance airplanes, the landing gear is supplied with a mechanism that allows it to be retracted into the plane when it is not in use. The retraction system may be manual, electric, hydraulic, or some combination of the three. Add seats, and you have the bare essentials of an airplane. To make the aircraft easy and safe to fly, however, you need a few basic instruments: an altimeter, which gives your height; an airspeed indicator, which tells you how fast the plane is going *with respect to the surrounding air;* a compass for orientation; and a few basic engine instruments, which keep you posted on the situation under the hood. Nothing more than that is needed to fly.

Once you start equipping an airplane beyond the minimum necessities for flight, there is practically no limit to the degree of sophistication you can attain or the amount of money you can spend. Only weight limitations and lack of space prevent a light airplane from carrying equipment that is equal in capability to that of an airliner and in the medium- and turbine-twin categories even those limitations can't stop you. With the frills removed, however, all airplanes are quite similar: thin, light aluminum shells of a shape now commonplace but one that eluded mankind for centuries − the shape that flies.

2.

HOW WINGS WORK

The simplified explanation usually given of lift — that air passing over the top of the airfoil must go farther than that passing beneath and therefore goes faster and, according to Bernoulli's theorem, exerts less pressure — is approximately accurate, but it is more an aid to visualization than an explanation. It does not help one to understand how a symmetrical airfoil generates lift, why an airplane with a cambered airfoil is capable of flying on its back, or why a very small increase in angle of attack produces quite a large increase in lift. A thorough understanding of these phenomena requires a thorough understanding of a phenomenon called *circulation,* and a complete explanation of that factor is beyond the scope of this discussion — it usually takes about 20 pages and a good deal of integral calculus in basic aerodynamics texts. It is comforting to reflect, however, that fluid dynamics is a field about which not everything is yet known, and that not even the most sophisticated aerodynamic theory fully explains the behavior of air flowing about a wing. Our own incomprehension, it should solace you to know, is therefore different in magnitude rather than in kind from that of more learned men.

Bernoulli's equation, as applied to the case of a wing moving through air — or, as science would have it, to air moving over a wing — requires that pressure decrease as velocity increases and vice versa. Everything immersed in the atmosphere is subjected to a steady "background" pressure, which depends on how much atmosphere is above the object — or, as we more usually say it, it ends upon height above sea level. At sea level, the background pressure is approximately 14.7 pounds per square inch. Since this pressure is everywhere, it is simply ignored in discussions of lift: a negative pressure or underpressure means not a pressure less than that of a vacuum, which would be impossible, but a pressure less than atmospheric. The pressure on various parts of the surface of a wing immersed in moving air is the sum at any point of two values: the dynamic pressure, or impact pressure, of air particles against the surface; and the static pressure, which is the local air pressure as it would be measured by a barometer vented to the surface of the wing.

The surface of the wing may conveniently be described as consisting of three regions. The first lies in the vicinity of the leading edge and extends aft a few percent of chord on both the upper and lower surfaces: this is an area in which the wing presents a fairly blunt surface to the oncoming air, and impact pressure relatively high. The maximum impact pressure occurs along a line called the *stagnation line,* which is the wing's continental divide: all air striking anywhere above the stagnation line passes above the wing, and all air striking below it passes under. Air molecules that strike the stagnation line itself are brought to a full stop and then depart one way or the other, hence the term "stagnation," meaning "standing still." The stagnation line moves downward and backward as the angle of attack is increased.

The next region consists of most of the upper and lower surfaces of the wing. Air flows over the wing surface at high speed, but the surface is nearly parallel to the direction of flow; thus impact pressure is small, and, because of the high speed of the air passing over the surface, the static pressure is also lowered from atmospheric. Both the upper and lower surfaces of the wing are therefore subjected to reduced, or *negative,* pressure. This is apparent in a fabric-winged biplane: both the upper and the lower surfaces tend to bulge outward between the ribs. The same phenomenon is demonstrated if you open the little side window, the clear-vision panel that is beside the pilot in many airplanes: there is an obvious suction, or underpressure, at the window, which is the result of air moving across it at high speed. At sufficiently high speed, a fabric wing or even a very thin-skinned metal one can explode because of the low pressure over the upper and lower surfaces. The third region lies in the immediate vicinity of the trailing edge, where there is a region of dead air, in which the pressure is slightly higher than ambient.

The sum of pressures on both surfaces of the wing is positive at moderate positive angles of attack. As the angle of attack is increased, the three regions continue to exist, but their location and magnitude change somewhat: at high angles of attack near the stall, the impact pressure at the underside of the leading edge is high, the static underpressure on the upper surface advances to the leading edge and becomes very pronounced, and the static underpressure on the lower surface is reduced, because the speed of flow over the undersurface is retarded.

The fact that the wing has a sharp trailing edge is very important for the production of lift. If the trailing edge were rounded like the leading edge, there would be a second stagnation line slightly ahead of the trailing edge on the upper surface, the mirror image of the leading-edge stagnation line. In this case (using a somewhat simplified approximation of the nature of moving air), the velocities and pressure at the backside of the wing would mirror those at the front, and the total force exerted on the wing by the air would be zero. It is because the sharp edge does not permit air to flow around it from the high-pressure region beneath the wing to the low-pressure region above it that the wing is able to produce a discontinuity in the airflow: this discontinuity takes the form of a slight twisting or rotating motion that is applied to the air in the vicinity of the airfoil − a twisting downward of the airflow through part of a circle (which is what the term "circulation" refers to). The force exerted in thus displacing air has as its equal and opposite reaction the weight of the airplane; the weight of the airplane is sustained by the circulatory movement imparted to the air by the wing, and the means whereby the wing and the air interact is the pressure variation called for by Bernoulli's theorem.

The air in contact with the surface of the wing is stationary with respect to the wing. This may be seen in the fact that fine dust on the surface of the wing is not blown away no matter how fast you fly. The molecules of air that come in contact with the wing skin are carried along with the wing by surface friction. At some distance away from the wing skin, however, the free-stream velocity prevails; it follows that in the vicinity of the wing, there is an area in which friction exists between layers of air moving at different speeds. This area is called the *boundary layer*. At the leading edge the boundary layer is extremely thin, and it grows thicker, as you might expect, as the air flows back along the wing and disturbances near the wing skin have a chance to propagate outward. The boundary layer may be of two kinds − *laminar* or *turbulent*. A laminar boundary layer is one in which all particles move in parallel paths, and there is no swirling or eddying at right angles to the wing surface. A laminar boundary layer is very thin − perhaps 1/16" thick − and the amount of air being carried along by the wing is small; the drag of a laminar boundary layer is therefore small, and it is for this reason that airfoils shaped so as to produce a laminar boundary layer over a large portion of the wing surface are called *low-drag* airfoils.

As it moves backward across the wing, a laminar boundary layer, which is a fragile thing, eventually breaks up into eddies and whorls, like water flowing over rocks. The spanwise line at which the breakup occurs is clearly defined, because there is a clear structural difference between a laminar boundary layer and a turbulent one: this line is called the *transition line*. Behind the transition line all flow is turbulent, and the boundary layer rapidly becomes much thicker; the drag therefore also increases sharply. Near the trailing edge of the wing the boundary layer becomes "detached," and a small amount of air from the wake of the wing flows *forward* along the wing surface, creating the region of dead air earlier referred to as the third region on the wing. This forward flow may be seen by the behavior of waterdrops on the ailerons of low-wing planes: fat drops will sit still on the upper surface, merely jiggle about, or even creep forward *against* the apparent airflow.

The locations of the stagnation, transition, and separation lines on the wing are determined by (among other things) airfoil shape, angle of attack, and airspeed. Given a wing of a certain airfoil shape supporting a certain weight, the behavior of flow around the wing depends simply upon the angle of attack. At high angles of attack, the lift is high, not merely because the stagnation line has moved slightly downward and backward along the bottom-front surface of the wing but also because it has moved backward in a region of sharp airfoil curvature in which small displacements in the stagnation line induce very rapid flow. At high angles of attack, the local velocities at the leading edge may be six or seven times the free-stream velocity; though the average velocity of air particles traversing the upper surface is not a great deal higher than that of those passing under the wing, the local velocities thus may be very much higher − hence the large increments in lift coefficient available from small increments in angle of attack.

Since lift is so highly dependent on the location of the stagnation line it is apparent why a symmetrical airfoil can produce lift and why an airfoil may produce lift while it is upside down if its (negative) angle of attack is great enough. As the angle of attack increases, the separation line moves forward, and the region of dead air on the upper surface of the wing increases in size. At high angles of attack quite a bit of forward airflow may be observed through tuft testing wings. At a certain angle − around 15 to 18 degrees for conventional types of wings − conditions at the leading edge become such that the flow detaches itself from the upper wing surface just after rounding the bend of the leading edge − rather as the tires of a car may leave the ground if you drive over a hump at high speed. The dead-air region moves forward to cover the entire upper surface, and the wing is stalled. It still produces considerable lift, but the lift is due to impact and is accompanied by a huge drag, making sustained flight impossible.

Some airfoils stall gently by a progressive forward movement of the separation line. In others, such as the familiar 23000 series, the stall is violent, and the separation line moves forward with a leap. In yet others — very thin airfoils with sharp leading edges, a type never used in light aircraft — a "bubble" of dead air appears behind the leading edge, but behind the bubble the separated flow reattaches itself and behaves as on other airfoils: on these sections the stall is complete when the advancing trailing- edge separation and the expanding bubble meet. Airfoils that stall abruptly, such as the 23000 series, may be made into wings — like the Beech Bonanza's — that stall gently, by twisting and tapering the wing so as to prevent its stalling all at once. A progressive development of the stall from the wing root outward is induced, providing aileron control and unstalled tips throughout most of the stalling process, which is spread over several degrees of angle of attack.

In order to delay substantially the stall in any wing, it is necessary to prevent the forward migration of the separation line and dead-air region, or region of reversed flow. Devices are used that appropriate some high-pressure flow from the underside of the wing and send it over the top, increasing the energy of the upper-surface flow and helping it to beat back the reverse flow. A leading-edge slat, for instance, takes a stream of air from the high-impact-pressure region just behind the stagnation point and directs it backward across the upper surface, preventing forward movement of the reverse-flow region until unusually high angles of attack are attained. A Krueger flap, which is found on the inner panels of Boeing jetliner wings and which consists of a fence that swings down beneath the leading edge, artificially moves the stagnation line downward, increasing the volume of air passing over the wing and therefore the energy of that flow. Drooped leading edges and enlarged leading-edge radii similarly act to help the upper-surface air to round the leading-edge curve without becoming separated. Less common but more effective is boundary-layer suction: by perforating the upper surface (or both surfaces) of the wing and subjecting the interior of the wing to a strong underpressure by means of a suction pump, it is possible to suck the turbulent boundary layer into the wing, reducing its thickness and the wing's drag, and to delay the stall by ingesting dead air before it moves too far forward along the wing. A little-known and probably never used device, but one that is of at least theoretical interest, is the rotating leading edge. It consists of a long cylinder, which is turned at high speed so that its upper side moves aft, causing the layer of still air at the bottom of the boundary layer to move in a direction favorable to lift and thus delaying separation.

As you will have gathered, the leading edge of the wing is where the action is: its shape and smoothness strongly affect lift, drag, and boundary-layer characteristics. The after portion of the wing, except for the importance of a sharp trailing edge, is less critical and may be considerably distorted without greatly affecting wing characteristics. No part of the wing is unimportant, however; even small changes in shape near the trailing edge produce tiny changes in pressure distribution near the leading edge. An interesting example of the interplay of bodies at a distance from one another in an airflow is a retractable landing gear. A low-wing airplane with retractable gear stalls at a lower speed with gear down than with gear retracted, because the gear protruding beneath the wing creates a pressure hump, displacing the stagnation line and increasing flow energy over the *top* of the wing − rather like a Krueger flap.

3.

FLAPS

There is a fundamental difference between the effects of *leading-edge* high-lift devices and *trailing-edge* devices. Leading-edge devices such as slots, slats, and drooped leading edges serve to delay the stall or to render it diffuse and innocuous enough so that a pilot can fly at low speeds and high angles of attack without feeling that he is at the edge of an abyss. These devices extend the lift-producing ability of the wing into a higher angle-of-attack range, and, since lift increases as the angle of attack increases, they permit the wing to generate more lift at a given speed. Fixed slots and movable slats — Stinsons and Swifts used the former, while the Helio has the latter — may permit considerable increases in lift coefficient by delaying the stall up to angles of attack of 25 or 30 degrees (15 to 18 degrees is normal for a naked wing). You pay for the gain with the inconvenience and the possibility of control problems of flying in an extremely nose-high attitude. Drooped leading edges, such as are offered on a number of STOL conversion kits available for standard aircraft, provide relatively small improvements in lift coefficient, but they improve the stability and continuity of airflow over the wing at low speeds and permit the pilot to use the extreme low-speed range of his airplane without much risk.

Trailing-edge flaps work differently. Instead of permitting the wing to attain a greater angle of attack without stalling, they permit it to attain higher lift coefficients at a given angle of attack: the maximum angle of attack obtainable with a deflected flap is usually somewhat smaller than the maximum angle obtainable with the plain wing. Leading- and trailing-edge devices can be combined to get the most out of the wing: this is the system used on most airliners and many STOL aircraft, but its high costs and structural inconvenience make it uncommon in light aircraft.

There are several ways of understanding what the flap is doing. You can visualize that the wing holds the airplane up by deflecting air downward and that the flap pushes more air farther down. More technically but no less correctly, the flap increases the *camber* of the wing. This term is often misused. Sometimes *camber* is used to denote the curvature of the upper or the lower surface of the wing: if the undersurface of a wing is partially concave, it is said to have "undercamber." This is a useful enough expression, but it has little to do with the real meaning of the term, which is *the curvature of the mean line* of an airfoil. The mean line is an imaginary line exactly halfway between the upper and lower surfaces. In a symmetrical airfoil, the mean line is straight: there is no camber. Airfoils are more commonly somewhat humped: the upper surface bulges upward more than the lower surface bulges downward; the mean line is curved, or cambered. Mean lines can have many different shapes: they may be shallow circular arcs, as in most laminar airfoil sections; all the curvature may lie at the front of the airfoil, with the rest forming a straight line, as in the 23000 series still used on most Beech aircraft; or they may be reflexed or ogival in shape.

The camber line should not be confused with the *chord line*, which is merely a straight line drawn from the leading edge to the trailing edge irrespective of the airfoil shape. Only in symmetrical airfoils do they coincide. If an airfoil shape is built around a chord line and the camber is varied, what happens? For a given angle of attack − zero degrees, say − the greater the camber, the greater the lift, up to a point. Within the range of shapes that are practical for airplane wings, lift can be said to increase along with camber. The trouble is that as you increase camber, you *reduce* the stalling angle of attack of the wing. Therefore, while increasing the camber increases the lift at a given angle of attack, doubling the camber does not double the maximum lift, although it does increase it noticeably. (Doubling the camber means doubling the maximum distance between the mean line and the chord line. Camber is often expressed by giving that distance as a percentage of airfoil chord: values of 0% to 4% of chord are the most common.) Thus, the purpose of trailing-edge flaps is to increase the camber and therefore to increase the maximum lift coefficient.

Why use a flap instead of just cambering the whole wing to begin with? As usual, a lot of factors vary at once: in this case lift is a function of camber, but so is drag. In order to reduce drag to a minimum at cruising speeds, designers select a camber that gives the best combination of lift and drag at cruising speed; this is invariably a rather flat camber compared to one that would give the highest maximum lift coefficient, which explains the need for a device that could vary the camber in flight. The importance of increased maximum lift coefficients arises from a similar argument: low maximum lift coefficients would be tolerable if there were sufficient wing area, but wing area produces drag, and, again, it is the cruising speed (for all but the most uncompromising STOL designs) that provides a target wing area. More wing area than the optimum is almost always used in light aircraft, because a small sacrifice in cruising speed is deemed more acceptable than the complication and cost of very sophisticated high-lift devices or the high landing speeds of jets.

Sometimes a wing has too much camber, either in order to compromise between cruising and climbing efficiency or because it was designed for cruising at a certain weight and the plane happens to be more lightly loaded. Some sailplanes are equipped with trailing-edge flaps that deflect upward as well as downward, and a few airplanes have ailerons that can be allowed to float upward together several degrees. The effect is to reduce both wing-profile drag and trim drag and to increase the cruising speed or top speed. It is unfortunate that this simple and useful function hasn't found its way into more airplanes. Camber that is continuously variable for speed and load would give efficiency a boost, and, more important, it would be fun to play with.

There are only three types of flap in current use, omitting outlandish anomalies such as the external-airfoil flap and the quaintly named Zap flap, which are infrequently encountered: the plain flap, the split flap, and the slotted flap. They all work similarly: by deflecting the trailing edge of the wing downward they increase the camber. All of them produce a large drag increase as they are deflected, but some produce more lift than others.

The plain flap is the simplest: it is like an aileron that deflects only downward around a hinge near its own leading edge; between it and the rest of the wing is a seal, and no exchange of air takes place between the upper and lower surfaces. It does nothing to increase the wing chord, and it suffers from both a high drag penalty at deflections beyond 20 degrees or so and a relatively low lift contribution − about a 30% increase at the most. Because of its poor performance the plain flap is mostly used in very low-cost aircraft, for which ease of construction is the overriding consideration, and usually more for steepening the glide than for reducing the landing speed. (Landing speed is a function of the square root of the lift coefficient: increasing the lift coefficient by 30% over half the wingspan reduces the landing speed only by 7% or so − from 50 mph, say, to 47 mph.)

The split flap shows up on the DC-3 and on the 300- and 400-series Cessna twins but practically nowhere else. It was in vogue during and after the war, but it then gave place to the slotted flap. The split type consists simply of a hinged portion of the wing's undersurface − usually about the last 20%, or 25% of chord − which swings down like a plain flap but leaves the upper surface in place right back to the trailing edge. Surprisingly, although it produces a large region of dead air behind the wing, the split flap produces somewhat less drag for a given lift coefficient than the plain flap. It also provides as much as 50% more lift than an unflapped section. Again, assuming a flap covering about half the wingspan, an airplane with a flaps-up stall speed of 50 mph will stall at 45 mph or so with a fully deflected split flap.

For the moment, the slotted flap has inherited the earth: almost every light airplane built today has it. Such a flap is characteristically hinged about pivot points just below the wing undersurface (on all the Cherokees, for instance) or set on rollers in curved tracks (as on all the high-wing Cessnas). The fixed pivot points have the advantage of simplicity of manufacture, while the tracks permit somewhat more efficient use by introducing what is called *Fowler action,* after Harlan Fowler, its pioneer. Fowler action involves moving the flap backward as well as deflecting it downward − rather a cheap way to immortalize one's name. In effect, it increases the wing area. The slotted flap is shaped like a small auxiliary wing, and, when it is deflected, the upper surface of the airplane wing extends back to form a lip that just slightly overhangs the leading edge of the flap. The shape of the slot entry and the relative positions of the lip and the flap determine the effectiveness of the whole installation: slotted flaps are sensitive to small changes in slot geometry, and they are carefully designed to conform to published experimental data.

The advantage of the slotted flap is that, by funneling some of the high-pressure, high-energy air passing below the wing into the converging slot and spitting it out along the upper surface of the flap, it delays flow separation over the flap, just as a leading-edge slat delays separation over the wing. The region of turbulent and reversed flow over the trailing edge of the flap is reduced in size, and more air gets deflected farther downward more efficiently. Slotted flaps accordingly provide smaller unwanted drag increments than plain and split flaps and rather higher maximum lift coefficients − up to about 2.8, or nearly

18

90% more than a plain airfoil. Again assuming a 50%-of-span flap and a flaps-up stall speed of 50 mph, a good slotted flap will bring the stalling speed down by 17% to around 42 mph.

Slotted flaps may be improved by Fowler action or by throwing in additional slots. In a double-slotted flap a leading-edge slat has been added to the flap airfoil — it is called a *fore flap,* or *vane,* in this case. Some Boeing airliners use triple-slotted flaps, in which the flap airfoil is equipped not only with a leading-edge vane but also with a trailing-edge slotted flap and considerable Fowler action as well. Such a flap may produce local lift coefficients as high as 3.5, a gain of 230% over the plain wing. Once you get into flaps of this degree of sophistication, wasting 50% of the span on ailerons is no longer acceptable. Bonanzas, Comanches, Mooneys, and some Cessnas — such as the Cardinal and Centurion — have slotted flats that extend over 70% or so of the trailing edge. Helio STOL aircraft use a narrow-span, wide-chord aileron as well as a small upper-surface spoiler to preserve good lateral control, while something like 80% of the trailing edge is given up to flaps. The Mitsubishi MU-2 goes farthest of all: deriving all its roll control from upper-surface spoilers, it uses full-span double-slotted flaps, a drooped leading edge, and propellers driving air backward over a large part of the wing. Lift coefficients 300% superior to those of a plain wing permit this highly sophisticated and efficient airplane to combine Learjet wing loadings with Queen Air landing speeds. Such a wing on our sample airplane that stalled at 50 mph flaps-up would permit a reduction in stalling speed to 28 mph. In the MU-2G, the small, elaborate wing pays off in cruising speed: 295 knots, as opposed to 260 or less for most comparable but less refined types.

You always pay some price for efficiency, and flaps exact their toll. Full-flap go-arounds may be difficult or impossible because of the drag, especially in low-powered singles and in twins with one engine shut down. The flap moves the wing's center of lift backward, producing what is called a *pitching moment* — in this case a tendency for the nose to go down. At the same time they alter the angle of airflow — called the *downwash angle* — over the horizontal tail. In highwing designs this change may cause the plane to pitch up rather than down, a potentially unpleasant effect since, as the flap goes down and drag comes up, you don't want your nose coming up as well. It is worth remembering, incidentally, that, within the green arc on the airspeed indicator, the plane will fly equally well with or without flaps but at different angles of attack, depending on the flap position. In other words, there is no reason to let the airplane sink when you raise the flaps, so long as you are not below the flaps-up stalling speed. If the airplane seems to sink when you raise the flaps, raise the nose accordingly, and you will find that your rate of climb will hold steady or even increase.

In short-field and soft-field takeoffs, flaps are used to lower the speed of liftoff and to lighten the load on the tires. It is sometimes said that the best short-field performance is to be had by keeping the flaps up until liftoff and then deflecting them suddenly to the takeoff setting (usually 10 to 15 degrees). The idea is that during the takeoff roll the flaps only produce drag and impair acceleration and that they should be saved until their lift-producing function is needed. This makes sense, especially for low-wing airplanes, which trap a lot of air in a draggy venturi between a deflected flap and the ground. The advisability of using the technique depends, however, on the aptitude of the individual pilot for doing several things at once. If you are so busy rotating and trying to set up best-angle-of-climb speed that you accidentally let the flaps down to full just as you lift off, you probably should not try it in the first place. For soft fields, however, it goes without saying that the flaps should be deflected from the beginning − even perhaps a little beyond the normal takeoff setting − and the airplane be kept rolling in order to avoid bogging down.

We have not yet seen the heyday of the high-lift device. Flaps are cheaper than engines, and more power and smaller wings are the only elegant means currently at hand of making planes go faster. At the moment, we keep getting more power, willy-nilly. Maybe someday somebody will try refinement for a change. Where there's life (lift?) − as Tarzan used always to say with most unsimian sententiousness − there's hope.

4.

STABILATORS

Stabilators — ever wonder how they work? If you're beyond the stage at which you take all the shapes and sizes on faith but are not quite up to designing your own (if you had to escape from Crete, for argument's sake), you might wonder how that wobbly slab manages to do the same job as the trusty old stabilizer-elevator-trim-tab arrangement and why some designers choose to use it rather than the conventional tail. Why, for instance, did Cessna, after getting rich on planes equipped with conventional horizontal tails, put a stabilator on the Cardinal — and why did costly catastrophe follow? Or why did Wing drop the Derringer's stabilator for a conventional setup after losing a test aircraft in the Pacific Ocean?

The principal aim of the aeronautical engineer is to reduce drag and weight. Designers also seek to reduce complexity — in the number of individual parts that must be manufactured and attached to one another. The stabilator meets these requirements: because its entire surface reacts effectively against the airstream whereas only a portion of the total area of a conventional tail does so, a stabilator may be smaller than a conventional tail and achieve the same given degree of control effectiveness, or it may give better effectiveness for a certain size. Not only can it be made smaller (and hence lighter and less draggy), but it may also be made with fewer parts. A well-designed stabilator therefore represents an improvement over the two- or three-part horizontal tail.

It consists simply of a single slab, usually cut away in the center to allow for the fuselage, with a small beam or tube carrying through the fuselage and holding the two sides of the stabilator rigidly in the same plane with respect to one another. The trailing edge, or a portion of it, is equipped with a narrow flap called an *antiservo tab,* which moves in the same direction as the rear half of the stabilator: "up stabilator" (trailing edge up, corresponding to "up elevator") produces "up tab." The whole thing is pivoted around its own 24%- or 25%-of-chord point, usually on bolts in brackets attached to the rear fuselage bulkhead. By pivoting about its own center of lift (which is always around 25% of chord), the stabilator is neutral in pitch: without the antiservo tab it would just sit any which way with respect to the airstream. The purpose of the tab is to make the stabilator line itself up with the airstream. By deflecting upward when the trailing edge of the surface rises, it pushes the trailing edge downward; deflecting downward against "down stabilator," it pushes the trailing edge of the stabilator back upward. In fact, the tab is always pushing one way or another, except when it is exactly aligned with the rest of the surface; the stabilator therefore tends always to return to and maintain this alignment. The stabilator-tab combination may be said to be stable: if it is pushed out of alignment, the antiservo tab pushes it back.

The same end could be achieved by pivoting the stabilator about a point ahead of 25% — say, at 5% or 10% of chord; then it would not need an antiservo tab and would simply weathercock into alignment with the airstream. There are two reasons why this is rarely done. One is to prevent flutter, the stabilator must be at least partially balanced in weight around its pivot points: the farther forward the pivot points, the more dead weight is needed to balance the surface. Furthermore, some sort of trim-adjustment tab is necessary in any airplane; as long as you are going to have a trim tab, why not use it as an antiservo tab — a matter of actuator geometry — and move the pivots as close to 25% of chord as possible, thus minimizing the required counterweight? This is the course that designers have adopted in the majority of cases.

Generally, the stabilator is actuated by a pushrod or cables attached to the carry-through member within the fuselage. The trim tab is also pushrod-actuated, and its antiservo motion is achieved simply by the location of the ends of the pushrod with respect to the tab hinge line and the stabilator pivot point, respectively. By moving the forward attachment of the tab-actuator rod up and down, or fore and aft, usually with a screw jack, the stabilator incidence (with respect to the wing) at which the tab aligns itself with the surface of the stabilator can be changed: in other words, the surface can be trimmed. This is easier to see than to describe, and a peek into the tail stinger of a Cherokee, Comanche, or Cardinal will reveal how the mechanism works. It is important to understand that the control feel of the stabilator arises from the antiservo tab, which, by deflecting more and more as the stabilator is deflected more and more, produces a stick force that increases in proportion to stick displacement. Stick force is zero at the trimmed stabilator angle, just as it is zero at the trimmed elevator angle on a conventional type of tail, and the airplane thus automatically seeks out the pitch attitude (angle of attack, in other words) at which the stick force — properly called the *stabilator hinge* moment — is zero.

While a conventional horizontal tall loses effectiveness at extreme deflections, because the stabilator portion works somewhat against the elevator (in nose-high slow flight the stabilizer is at a positive angle of attack, producing lift, which must be overcome by a large deflection in order to produce a net down load), the stabilator actually becomes increasingly effective at large angles of attack, because its antiservo tab acts like a wing's trailing-edge flap, increasing the maximum lift available from the surface. By drooping the leading edge of the normally symmetrical stabilator airfoil section Beech's engineers made the Musketeer stabilator especially effective: its maximum lift coefficient may be as great as the wing's.

Some people are suspicious of stabilators. The Cardinal had bad luck with its stabilators, and it is still hard for some pilots to land the plane consistently, especially with full flaps. The original reason for using a stabilator on the Cardinal was that only a stabilator could produce sufficient pitch control to handle the plane's extreme forward CG location and at the same time be reasonably light in weight. The forward CG in turn was the price of placing the pilot so that he could see out from under Cessna's traditional high wing. There were a number of landing accidents on the early Cardinals, followed by a recall of all aircraft and the installation of a leading-edge slot on the inner portion of the stabilator. According to Cessna, the mod was intended not to smooth out the landing but to eliminate a forward-pitching tendency that originally existed when the plane was slipped with full flaps. In any case the stabilator-equipped Cessna has a different landing feel from the others. Whereas a conventional horizontal tail becomes decreasingly effective at high angles of attack and the control movements used during flare are large and rough in comparison to those used in normal flight, the stabilator seems to have nearly a constant feel throughout the full angle-of-attack range. The result is that perfect landings may be made consistently by a gradual, steady increase of back pressure during the flare, while the large yoke movements customary in other Cessnas produce porpoising in the Cardinal. Despite its reputation the Cardinal is, if anything, *easier* to land well than some other Cessnas, but it was slow in gaining customer acceptance.

Airplanes have occasionally shed stabilators at high speeds. Since they consist of a large moving mass that is only partially counterbalanced, stabilators must be carefully analyzed for dangerous resonant frequencies and possible proneness to flutter. Some of the mass balance must sometimes be moved out to the tips of the surface to prevent flexural flutter between the overbalanced center and the unbalanced tips. Especially in twins, the tips of the stabilator may sometimes be loaded differently because of slipstream differences between the two engines, leading to flexing and vibration in the central torque box; this may have been the cause of the Derringer accident. Furthermore, FAA horizontal-tail load requirements, developed for conventional tails, are just barely adequate for all-flying ones, and designers who work from Part 23, Appendix A − the Federal guidelines for light-aircraft design − must give particular attention to the strength and rigidity of all-flying tails. To further increase the problems, comparatively little technical data is available on stabilator design for light aircraft. Nevertheless, the advantages of the all-flying tail over the half-flying type have persuaded many manufacturers to adopt it. Most jet fighters are equipped with this type of tail (the F-4, F-105, and F-111 come to mind immediately), as are a large number of single- and twin-engine light aircraft. The proneness of a few stabilator-equipped aircraft to "hunt" − that is, for the nose to bob slowly up and down − in cruising flight is usually due to friction or play somewhere in the tab-actuator circuit. Like any other novelty, stabilators get blamed for everything that goes wrong, but they're here to stay. If you don't like them, you had better get used to them so that you can adapt when the all-flying *vertical* tail arrives.

5.

PROPELLERS

The propeller is a wing: it produces lift and lifts the plane forward through the air. A helicopter rotor is a glorified propeller, or a rotating wing. The borderline between the genres is patrolled by tilt-wing aircraft, whose huge, paddle-bladed props lift the ship vertically off the ground and then tip forward to propel it through the air. Since the propeller is a wing, its functions can be thought of in terms of lift, drag, angle of attack, and so on. Like a wing, a propeller may be stalled; like a wing, it experiences profile and induced drag and trails tip vortices.

The simplest propellers are fixed-pitch. Pitch is the angle between the blade and the imaginary propeller disk at any point along the blade. This angle determines the angle of attack of that blade element at a given combination of engine rpm and airspeed. The pitch angle varies along the blade, making it appear twisted, because at a given forward speed, the angle at which any blade element encounters the air is the sum of two motions: forward speed and rotational velocity. Since the tips of the propeller move a greater distance through the air with each revolution than do the roots, they obviously move at a higher speed. The roots are steeply angled with respect to the prop disk, because the principal component of their motion through the air is the forward speed of the airplane; the tips, on the other hand, may be moving at 400 knots around the hub and only 100 knots forward: their pitch angle is thus closer to the plane of the propeller. Since blade angle varies along the radius of the prop, pitch is often defined (for convenience) as the number of inches that the prop would move forward per revolution if the blade elements were maintained at a certain angle of attack.

The weakness of fixed-pitch props is that they make the engine output dependent upon airspeed. The faster the plane is moving forward, the lower is the angle of attack of the blades with respect to the air passing through the prop disk. The lower their drag, the faster the engine revs. You are obliged to reduce throttle as you pick up speed in order to avoid passing the engine redline. As you slow down, on the other hand, the angle of attack of the blades increases, and you have to add power to overcome the increased drag of the blades. Power output is a function of manifold pressure (mp) and rpm. Since maximum mp is limited on unsupercharged powerplants by atmospheric pressure, maximum power can be developed only by increasing rpm. Since maximum rpm depends on airspeed, an engine driving a fixed-pitch propeller may develop its rated power at only one forward speed or over a small range of speeds. That is why climb props, cruise props, and speed props have been developed. A climb prop develops maximum continuous engine power at approximately the best-rate-of-climb speed; a cruise prop does the same at cruise speed; and a *speed,* or *racing,* prop, at top speed. An airplane with a speed prop will have a higher top speed than one with a climb prop, though they be otherwise identical: the engine driving the climb prop will not develop its maximum power at the airplane's maximum speed. On the other hand, an airplane with a speed prop will be deficient in takeoff and climb. In practice there is a fairly wide range of speeds at which these power losses are less than 10%, which makes it possible and customary to equip planes with compromise props that give satisfactory performance in both climb and cruise. The benefits of being able to change prop pitch in flight have always been apparent, and primitive variable-pitch props were first developed with two positions-one for climb and one for speed. Later types had continuously controllable pitch, but rpm at a given pitch setting was still dependent on airspeed.

The modern constant-speed propeller is a convenient adoption of the controllable-pitch mechanism. Instead of being directly controlled from the cockpit, pitch is automatically regulated by an engine governor, and the pilot selects engine rpm from the cockpit. High-pressure engine oil supplied by a booster pump in the governor holds the prop blades at the appropriate pitch setting for the rpm chosen by the pilot. If you increase airspeed or mp, a flyweight control in the governor responds to an incipient rise in rpm by increasing oil pressure to the prop, which moves the blades to a coarser pitch setting-that is, a higher angle of attack-to increase blade drag and restore the selected rpm. If oil pressure from the governor is reduced either by the pilot or by a mechanical failure, the centrifugal forces within the prop normally move the blades to fine pitch; some props are equipped with counterweights to assist in this operation. The fully forward position of the cockpit pitch control is called *flat pitch,* or *fine pitch,* and roughly corresponds to a climb prop. The blade angle of attack is low; drag is low; rpm is consequently high. If you pull the control back and shift the blades toward the *coarse-pitch* position, their angle of attack is increased; their drag rises; and rpm drops.

A given percent of power may be produced by a variety of combinations of mp and rpm, just as sufficient lift to stay airborne may be produced by several combinations of airspeed and angle of attack. Once an rpm setting has been selected, pitch alters automatically with mp, and the throttle acts as a direct thrust control. In practice, the best operating efficiency is achieved by maintaining low rpm (consistent with handbook limitations), because the prop is most efficient at lower rotational speeds and because less fuel is wasted in turning the engine itself at 1,900 rpm, say, than at 2,400. Leaving the prop in coarse pitch on final approach, however, reduces the power available for a go-around, since you can develop maximum mp if you firewall the throttle but not maximum rpm. Furthermore, the fine-pitch setting has a braking effect at approach speeds (because the angle of attack of certain blade parts may be negative) if the throttle is reduced to idle. For both of these reasons, "prop to fine pitch" is an important part of the prelanding cockpit check.

Pilots often long for an extra prop blade or two. The principal gain lies in compactness and possibly in noise reduction − not in performance. A three- or four-bladed prop is simply a two-bladed one squeezed into a smaller disk, just as a biplane is a monoplane squeezed into a shorter wingspan. Long, slender blades, and as few of them as possible, are a positive asset in takeoff and climb and do no harm at lightplane cruise speeds. The multibladed, paddlelike propellers on very large engines are a compromise between weight, ground clearance, tip velocity, and the need to absorb the engine power output. They are often less efficient in takeoff and climb than two-bladed lightplane props.

Pusher propellers are identical in operation to tractor props: the engine is simply moved to the other side of the propeller. Some performance advantage is to be expected from removing the fuselage or nacelle from a position within the propeller slipstream, where the velocity is higher than that of the surrounding air. On the other hand, a pusher propeller operates in air that has been disturbed by the passage of the fuselage or nacelle before it. As it turns out, the gains and losses very nearly cancel out one another, and pusher configurations are not demonstrably superior to tractor types; tractor designs have other advantages, however − cooling, weight distribution, ground clearance − that account for the preponderance of tractor-engine aircraft on the market and in the history books. Partisans of the pusher layout often advance the Cessna Skymaster as evidence for their case, since it performs better with its front engine feathered than with the rear one feathered. This is a special case, however; the battleship-like stern of the Mixmaster is so blunt that, without the low pressure generated by the after propeller to suck air around its corners, flow separation occurs and drag rises sharply. This (to show how an author's favorite moral may be squeezed from even the driest matter) is a good example of the multiplicity of factors involved in all things aerodynamical and of the usual inadequacy of simple explanations.

6.

PROP AND THROTTLE

"Revolutions per minute" (RPM) hardly needs explaining: the term is completely unambiguous. "Manifold pressure in inches of mercury" (mp) is a little more obscure. To start with, how did mercury get into the engine, and, more subtly, why is there pressure instead of suction in the manifold (or does it refer to the exhaust manifold)? Furthermore, by what magic can you be safe and sure with any engine as long as you obey the "square law" and keep the number of inches of mercury in the manifold equal to the number of hundreds of turns per minute of the crankshaft? Why does departing from the "square" in a certain direction sometimes entail a sudden unexpected landing? And what, for that matter, is a manifold?

To begin with, manifolds are the pipes through which the fuel-air mixture is sent to the cylinders (intake manifold) or the exhaust is carried away from the cylinders (exhaust manifold). The term *manifold pressure* refers to the absolute air pressure (that is, measured from zero, not from one atmosphere, as tire pressure is measured) inside the *intake* manifold. A clue to the meaning of mp is found in the altimeter. The range of barometric pressures displayed in the sensitive altimeter-adjustment window more or less coincides with the upper limit of the mp scale on nonsupercharged engines. As a matter of fact, both refer to the same thing: air pressure. The altimeter reads the local atmospheric pressure, and the mp gauge reads the pressure inside the intake manifold. Because the intake-manifold pressure is supplied directly from the outside air, the mp can never exceed the atmospheric pressure (except, again, in supercharged engines). That is why the maximum available mp drops lower and lower as you gain altitude.

Let's start with the basics. Pressure is measured in English-speaking countries in terms of inches of mercury, which refers to the pressure exerted at the bottom of a tube filled with mercury (Hg). The standard pressure of air at sea level − that is, the pressure exerted at the bottom of a filled with air to the height of the atmosphere − is equal to the pressure at the bottom of a column of mercury 29.92" tall. Barometric pressure varies from day to day, but it always remains in the vicinity of 30". Without the altimeter setting the altimeter cannot be used very accurately.

Consider how fuel is fed into a conventional piston engine: suction (that is, pressure lower than atmospheric pressure) is produced in the intake manifold by the downward travel of the piston in the cylinder, which draws air out of the manifold into the cylinder. The suction in the manifold in turn draws air in through the carburetor. The carburetor is equipped with a venturi throat, like those mounted outside some airplanes, to provide suction for instruments. The carburetor venturi also provides suction, which draws fuel out of the carburetor float chamber and allows it to mix with the inrushing air in the form of a fine spray or vapor. (It works just like the old-fashioned preaerosol bug sprayers, which sucked liquid up out of the can and turned it into a fine spray; paint-spray guns also work the same way.) The amount of the fuel-air mixture that reaches the cylinder is controlled by a throttle valve in the carburetor throat between the venturi and the manifold. When this valve is wide open, air rushes rapidly into the manifold as the piston descends on its intake stroke. The pressure in the manifold remains close to atmospheric pressure, because nothing is preventing the outside air from entering the manifold and cylinder and filling up the void. An open throttle thus corresponds to a manifold pressure close to atmospheric — around 28" or 29" of mercury. Because of obstructions inside even an open carburetor the manifold never quite manages to fill itself up with air as fast as the pistons pump it away: hence the small difference — 1" or so — between atmospheric pressure and the maximum manifold pressure available without supercharging. If you close the throttle, on the other hand, you severely restrict the passage of air through the carburetor: less air can enter the manifold during the intake stroke, and the pressure within the manifold remains lower. Air simply can't get in quickly enough to maintain atmospheric pressure. A low manifold pressure thus corresponds to a low throttle and to a low power setting. The same principles hold true for fuel-injection engines, in which fuel flow is metered according to the amount of pressure in the intake manifold.

Most pilots know that it is hazardous to reduce rpm beyond a certain point while keeping mp above a certain point. The situation corresponds to that of a car driving up a steep hill in a high gear: the engine makes comparatively few revolutions, but each is called upon by the open throttle to do a lot of work. The pressure inside the cylinder, forcing the piston downward on the power stroke, is high; if it is high enough, it leads to overheating and knock. Knocking heats the cylinder head beyond its safe limits, and the piston is subjected to hammerlike blows from the exploding fuel-air mixture rather than the smooth, even pressure increase that occurs with normal ignition. Automobile engines seem to take a bit of knocking in stride, since they operate at comparatively low power settings, but airplane engines do not: they are designed to run closer to their maximum potential, and they cannot tolerate knock for very long — more than a minute, say.

Every engine manufacturer publishes in the engine operator's handbook a chart for each of his engine models showing the permissible range of rpm against mp to prevent knock. Using this chart or committing it to memory is a more rational procedure than squaring mp and rpm. It can also pay off in reduced fuel consumption, since rpms cost money. Consider an idling car. In order to increase the engine rpm, you step on the gas. The same is true of an airplane. Some fuel is required merely to turn over the engine, and the faster you want it to turn over, the more fuel you have to feed it to overcome the running friction of the engine. Fuel can be saved, therefore, by bringing rpm to the minimum and mp to the maximum for a given power setting. If you look at the manufacturer's operating recommendations, you will find that considerable departure from the square is possible. For instance, in certain 260-hp Continental engines, 22" of mercury and 1,900 rpm is a safe — and miserly — power setting. Low rpm also reduces total engine hours and increases tbo (time between overhauls), because the engine hour meter is really a rev counter that converts revolutions to hours by assuming that the engine will always be turning a certain rpm − 2,566 in most Cessnas, 2,053 in Bonanzas up to 1959, and so on.

It is important to stay within the operating limitations of the engine; if you are going to cruise at 1,900/22 and you have been climbing at 2,500/25, you cannot reduce rpm first and mp afterward, since you would then be running briefly at 1,900/25 − an unsafe setting. This is the reason for the basic rule: to reduce power, reduce manifold pressure first and rpm after; to increase power, increase rpm first and mp after. This also explains the P in GUMP, the before-landing check (gas, undercarriage, mixture, prop). Suppose that you have been cruising at 1,800/20, you have descended to land, and you forget to move the prop control to flat-pitch (high rpm). You're on final, and for some reason you have to go around. You firewall the throttle and get 1,800/29 − for as long as the engine lasts.

Fixed-pitch propellers are matched to the engines in such a way that they regulate mp by offering a resistance to the engine proportional to rpm. Except by overheating the entire engine, it is difficult to induce knock in an engine driving a fixed-pitch prop. Having separate rpm and mp controls, however, increases the risk to the engine by increasing the number of things that the pilot can do wrong. On the other hand, the two controls, properly understood, offer better performance, better fuel economy, and the possibility of more flight hours between overhauls.

7.

THE INSTRUMENT PANEL

Behind the quiet, unassuming exterior of an instrument panel lies a quiet, assuming interior. Not much is going on back there, but it is interesting and helpful to know exactly what is causing all those proudly erect or morbidly sagging needles to assume the attitudes they assume, especially since the instruments are the vital bearers of news as to how the airplane is getting on.

Most of the basic instruments are operated by electricity or by pressure. The gyro indicators work by capitalizing on weird properties of spinning masses. The remaining exceptions − a few temperature-measuring instruments − operate by harnessing the expansion and contraction of a fluid, thermometer-fashion, although in a sense they are really pressure measurers.

Four of the key instruments measure air pressure: the altimeter, the airspeed indicator, the vertical speed indicator, and the manifold pressure gauge. Essentially, they contain the same sort of mechanism, a flexible capsule that is hooked up to various gears and levers so that its expansion or contraction in response to inside-outside pressure differences turns pointers on the front of the instrument. In the altimeter and manifold-pressure gauge the pressures being compared are the static air pressure around the airplane and the pressure of a trapped volume of nitrogen gas within the capsule. To measure airspeed, static pressure is compared with the dynamic pressure of oncoming air ramming into a hole that is faced forward to receive it.

The vertical-speed indicator (VSI) is a variation on the altimeter. It has a tiny hole in its capsule that allows air to bleed to the case and eventually to neutralize the pressure difference created by climbing and descending. Because this hole is minute, the neutralization takes place slowly so that a change of altitude registers as a temporary needle movement in proportion to the rapidity of the change. Although this kind of VSI is old-fashioned and subject to lags and errors, it is still found in most light aircraft. It is gradually being replaced by the instantaneous vertical-speed indicator (IVSI), which contains not only a leaky altimeter but a small accelerometer to give the needle immediate notice of upward and downward motion, after which the normal mechanism takes over.

The altimeter and the manifold-pressure gauge are really the same instrument: the altimeter is simply geared to produce more needle movement for a given change of pressure. Manifold pressure is the absolute pressure inside the engine intake manifold. Many people are confused by the term *pressure,* since it is always (except in boosted engines) a pressure lower than the ambient pressure. The engine acts like a vacuum pump, sucking air out of its intake manifold. Air is admitted to the manifold through the throttle. When the throttle is wide open, the manifold pressure is closest to ambient. As the throttle moves toward the closed position, it restricts flow and thus causes the pressure within the manifold lessen in comparison with that outside. (This explains why the highest manifold pressure available without turbocharging is always around 30" of mercury − the same as sea level atmospheric pressure.)

Should you suffer an a altimeter failure during an IFR flight, you can get approximate altitude information from the manifold pressure gauge by fire-walling the throttle. Atmospheric pressure drops about 1" of mercury per 1,000 feet of altitude, so your height above sea level would be equal to about 1,000 times the difference between full-throttle manifold pressure and 28" (the 2" adjustment from 30" is for intake manifold losses, which may be as little as 1 1/2" at 2,000 rpm or 1 1/2" to 2" at 2,500 rpm). This is not the most accurate way of measuring altitude, but it is better than touch.

The manifold-pressure gauge is part of a closed system, as it is vented directly to the pressure that it measures. The altimeter and the airspeed indicator, on the other hand, are more complex in that they depend on ambient pressure, which is not easily measured from a moving vehicle.

The usual method is to take the static pressure along both sides of the airplane (to cancel yawing effects) from the region of "dead" air that is carried along close to the skin. This air is never altogether dead, however, and even if it were, its pressure would not necessarily be the same as the ambient pressure at a distance from the aircraft. The static system is therefore subject to a *position error,* which manufacturers try to use, within legally required tolerances, to their advantage. A static-port position that raised the stalling IAS while lowering the cruising IAS would call for further testing; one that minimized stall speed while exaggerating high speed would probably be considered close enough. A few owner's handbooks give tables of calibrated airspeeds (CAS) for the low end, where errors are usually hardest to eliminate. If the static ports or lines freeze or otherwise block, you have to switch to alternate static air. Lacking such equipment, the trick is to break the glass on the front of the VSI, thus venting the whole static system to the cabin. (You break the VSI glass because it is the most expendable of the static-related instruments; in breaking the glass you might damage the instrument itself.) If you have a system with an alternate static source, it is a good idea to check it in flight and note the airspeed and altitude errors that it introduces (the alternate system is almost always less accurate than the primary and never more so). If the errors are large, they should be noted or memorized: they are usually to the worse side for an instrument approach − the airspeed indicator and altimeter both read high. Do not confuse the static pressure with the suction provided by venturis mounted on the sides of some airplanes that lack vacuum pumps. The venturi generates suction passively as the upper surface of a wing does, and the suction is used to drive gyro instruments. It is rare today for an airplane to have gyro instruments and not to have a vacuum pump.

The air-pressure-measuring instruments are delicately constructed to measure delicate forces. Other pressures − oil, fuel, hydraulic, and the like − are measured with a hollow, curved metal device called a *Bourdon tube.* One end is open to the pressure source, and the other is closed and hooked to a meter movement. Under pressure, the tube tends to straighten, and the motion turns a needle. Oil, fuel, and hydraulic fluid normally don't reach these instruments unless the Bourdon tube springs a leak, since a volume of air is generally trapped in the pressure line when the system is first hooked up. The air transmits pressure just as does any fluid, and neither the instrument nor the pilot ever knows the difference. Some panel-mounted outside-air-temperature gauges and some of the older oil-temperature gauges in three-needle engine clusters also use a Bourdon-tube mechanism in which the driving medium is a trapped fluid flowing from a coil or bulb sensor through a capillary tube to the instrument. As the sensor is heated, the expansion of the fluid is felt by the panel instrument as a pressure and is registered accordingly. The dial face is calibrated in degrees, however, and read as a temperature.

Temperature is more commonly measured electronically, either by a thermocouple, also known as a thermel, or by a resistor whose properties vary with temperature. A thermel has a probe in which two dissimilar metals are in contact. When they are heated, the metals produce a tiny voltage that is sent back through wires of carefully controlled resistance to a sensitive galvanometer. What is read is a voltage, but the dial is again calibrated in degrees of temperature. Thermels have two wires that are matched to the instruments and should never be cut or altered. Resistance devices are recognizable in that they have only one wire, like the water-temperature bulb on a car. In the end they too control a voltage that is read as a temperature on a panel galvanometer. Cylinder-head, exhaust-gas, oil, carburetor-inlet, and turbine-inlet temperatures are usually measured electrically.

One basic instrument that measures neither pressure nor temperature is the tachometer. The garden-variety tach is mechanical and similar to an automotive speedometer. It is driven by a flexible cable that transmits engine rpm or some geared variant of it to a case in which a small magnet spins within a small aluminum cup. The moving magnetic field sets up electrical currents in the cup; the currents have their own magnetism and interact with the spinning magnetic field to tug the cup against a fine spring. The movement of the cup is proportional to the current strength, which in turn is proportional to the speed of the spinning magnet; the cup moves the pointer. Another type of tach is electronic. It uses a capacitor, which, like a small battery, is charged whenever a pulse arrives from a magneto to indicate that a plug has fired. The capacitor discharges at a steady rate, its voltage dropping as it discharges and rising as it is charged. The faster the engine turns over, the more frequently the capacitor is charged and the higher its voltage; this voltage drives a panel meter that is calibrated in revolutions per minute.

The compass is, of course, operated by the earth's magnetic field and is at once the most and the least useful of instruments. It eventually tells you which way you are going but suffers from embarrassing nervous disorders in the process. Several types of compass are more stable than the small liquid ones: some are panel-mounted, and others − remote compasses − are kept at a distance from iron or steel masses and from electronically induced magnetic fields that can cause confusing deviations. Remote compasses use a synchronous electric-drive system to slave the panel indicator to the compass. Yet another type, the flux valve, measures tiny currents induced by a coil that turns with the airplane in the earth's magnetic field. Flux detectors are free of the turning errors and lags of normal magnetic compasses and are sometimes used as primary directional references in slaved-gyro and autopilot systems. For all their advantages they are still rare in small planes.

Gyro instruments depend on the puzzling desire of rotating masses to resist movement of their axes — a phenomenon with which everyone is familiar but of which few people have any intuitive understanding. It is sufficient for our purposes to understand that, when the mass of the gyro is in rapid rotation, it tries to keep pointing the same way in space.

Artificial horizons use gyros on vertical axes, while directional gyros have their rotor axes in the horizontal plane. Turn coordinators have tilted gyros. As the airplane moves, so does the instrument case, but the gyro rotor and its immediate mounting remain stationary with respect to the earth. The instrument can thus provide a symbolic picture consistent with what the pilot would be seeing outside his windows. For orientation a symbol of the airplane is affixed to the front of the instrument case. Inconsistencies can create problems of interpretation. Some old attitude gyros had an outer-world display with the sky at the bottom and the ground at the top: when the airplane climbed, the gyro showed what would nowadays be interpreted as a dive. On the other hand, some recent attitude gyros have a display in which the outer world moves with the real airplane so that the horizon remains horizontal with respect to the pilot as the airplane bug banks with respect to it. In terms of the actual horizon, therefore, the airplane bug banks twice as far as the real airplane. This outside-looking-in display annoys some people, because the picture that it shows does not correspond to what one sees when looking up after breaking out of clouds on an approach to minimums. They say that the effect is confusing. Once one gets used to this type of display, it is no more difficult to interpret than another, and some find it easier.

Instrument design involves not only the mechanical problems of accurately and reliably transmitting information but also a human problem of making the information easy to understand. As instrument design is refined, questions of how best to display data to a busy, perhaps alarmed pilot are becoming increasingly important. The altimeter is one example of an instrument that seems easy to read but sometimes permits or induces fatal mistakes. The problem is not only how to make the information clear but how to make it unambiguous, so that even a pilot who has a preconceived notion that he is at 14,000 feet cannot misconstrue an altimeter that reads 4,000 feet. The present system of indicating increments of 10,000 feet by hour hands and striped flags is apparently unsatisfactory. Other instruments, such as oil-pressure, fuel-pressure, or cylinder-head temperature gauges, do not command much attention, and it would take most pilots a bit of good luck to notice a drop in oil pressure when it happened, rather than some indefinite number of minutes, hours, or days later. Annunciator lights are one solution to this problem, though they involve a lot of additional costly electronic circuitry and, like other add-ons, are one more thing that can go wrong. Many military aircraft use vertical-scale instruments that can be arranged so that all related indicator needles form a straight line under normal conditions. The pilot does not have to read each instrument separately but need only verify with a glance that the lubber line is straight.

Panel instruments are badly in need of modernization. The classical circular displays waste space, are often not relevant to the nature of the information being presented, and are poorly suited to emphasizing vital information. If you consider the confusion that invariably besets you for the first hour or so in an unfamiliar airplane, you may infer that all is not as it should be. Vital information about the attitude and position of the airplane should be gathered in a single display in front of the pilot. Engine and fuel information should be near it, arranged in such a way that a brief glance tells the pilot all he needs to know. Navigation display and radio-frequency selectors, clearly labeled, might be grouped to the right near the center of the panel, as they commonly are today. A head-up display of angle of attack should be placed above the panel, slightly to the left of the pilot's line of vision, and focused at infinity so that he can monitor this most important piece of approach information while looking at the runway.

There are compelling reasons why an overhaul of the World War II method of displaying flight information is unlikely. For one thing, "new" instruments are merely remanufactured old ones. Instrument manufacturers are interested in the retrofit market, and all existing panels are riddled with standard round holes; nothing short of complete replacement of the panel would do for a thorough reform of instrument design, and such a modification is beyond the budgets of most light-airplane owners, who, after all, have been doing fine for years with the old instruments. Manufacturers are unlikely to sink a lot of money into a radical concept that would be usable mainly in newly manufactured aircraft — and then only if their builders could be persuaded to take a leap into the unknown. Such leaps are rare in this business. The most radical work now being done in the field of instrumentation is in sophisticated flight directors and automatic flight-control systems for heavy business aircraft — mostly turboprops and jets. Some of the new situation displays and annunciators make inspired use of complicated electronics to simplify the pilot's job; by comparison, the majority of general-aviation panel equipment is barely emerging from the audiovisual Stone Age.

8.

BEING KIND TO ENGINES

Engines never complain, at least not to people who are insensitive to the sounds of metal in the throes of ultimate strain. The reciprocating engine has become man's best friend by literally working itself to death with never a word of backtalk. Some people work engines without a care for their feelings and get unfailing performance – up to the moment that the crank seizes. In airplanes that moment must be postponed as long as possible, for what is mere inconvenience in a car or truck can be substance for a movie script when it happens in an airplane. Manufacturers used to be dogmatic: "Do it," they said, without explaining why. But pilots like to think of themselves as enlightened, intelligent, and capable of deciding when to lean a mixture or open a cowl flap. By way of response manufacturers today expend some effort to explain why they say what they say in the manuals. Although they have become commendably open in their approach to engine operation, suspicions persist that most of the doctrine is aimed at protecting the pilot from himself. Are the usual operating-manual directives needlessly conservative and lacking in faith in pilots? Are we blindly following procedures that, although safe, are inefficient?

In the discussion that follows, all advice must be qualified with the familiar words, "When in doubt, consult the owner's manual." You may find opinions here that are at odds with what you've read or been told elsewhere. If you want the peace of mind that comes with being sure of yourself, ask a mechanic, ask an experienced chief pilot, talk to an FBO, and read texts on engines. When you are satisfied that you understand what is happening inside the engine during a flight, you can begin to operate it with reason instead of with blind obedience. There must be limits to any review of engine operation, and this discussion is confined to modern, opposed-piston, normally aspirated (as opposed to turbocharged) engines.

Engines deserve the attention and maybe even the affection of every pilot. It is a curious attribute of aircraft engines that they will work longest if they are worked steadily. A recent Lycoming reminder that engines deteriorate if they are allowed to stand for any length of time without running can be seen in practice at Piper's Vero Beach factory, where every Saturday morning is devoted to running up the engines in the dozens of parked airplanes for 45 minutes to heat the oil and drive away internal moisture. The best way to heat an engine, however, is to fly with it. Ground run-ups are better than nothing, but a full-power takeoff, a short cruise, and maybe a few touch-and-go's will not only keep the pilot sharp but will ensure that the engine will live to see its predicted life-span. Lycoming claims that 20 to 30 hours of use a month is the minimum necessary to attain the full service life. The company further advises that, if you must use ground run-ups, the oil temperature should be brought up to at least 165 F. Flying the aircraft will easily accomplish that and will further assist in driving away accumulated water by providing a crankcase-ventilating airflow.

The airframe manufacturer has provided the cleverest of traps for water and dirt in the fuel system, placing filters and drains at low spots along fuel lines and in tanks. Water is sometimes uncooperative, however: if you suspect that there is some residue in your fuel, it is best to drain more than the customary thimble's worth of gas; set aside that hollow screwdriver and get yourself a good-sized bottle. Should water get past filters, traps, and sumps and into your carburetor, it will eventually be sprayed into the venturi airflow to be mixed with air as if it were fuel. Water does not approach the volatility of gasoline nor will it burn, so the engine will simply shut down abruptly or run very roughly until heat boils the water off. If large quantities of water manage to find their way in, you are in for bigger trouble.

Injected engines have special problems if water penetrates the fuel system's defenses and finds its way into the fine fuel lines and metering-jet orifices that distribute fuel to each individual cylinder. If a dose of water is injected into only one cylinder, it will misfire, and the piston, crankshaft, or connecting rod will have to absorb the out-of-balance shocks.

Before actually starting up, walk to the front of the plane and address your engine head-on. The propeller that you see in front of you should be free from imbalance or runout. Either condition is easily misinterpreted as a roughness in the engine itself. If it is allowed to go unchecked, it will contribute to engine wear by creating excessive bearing loads and vibration. Check the air filter if it is accessible. Never run the engine with the filter removed: airplanes kick up so much loose dust that you'll almost certainly get some into the upper cylinder area, where just a few seconds of abrasion will wreck the ring seal. You can kiss your compression and your oil good-bye if that happens.

Peek inside the cowl and look for blockages or breaks in the cooling fins. Cylinders that have lost a certain percentage of their cooling-fin area have to be replaced, but baked-on mud or debris between fins can be cleaned easily. The fins increase the surface area of the cylinder, thereby offering a greater cooling surface to the passing air. If part of that surface becomes blocked, the designed-in cooling capacity of the engine may be reduced.

Are all the aluminum baffles there and mounted properly? Is the proper winter- or summer-cooling kit installed? Some airplanes need an extra baffle when the ambient air temperature drops below a certain point. Without the winter kit the engine may run too cool, which is not good. Check the baffles for cracking: they are generally fashioned of a light-gauge aluminum that succumbs to vibration more easily than the stouter parts do. Never remove baffles from an engine or fly with a baffle missing on the assumption that the engine will run even cooler without it. With the exception of certain temperature-regulating baffles that block off cylinder-fin area, most baffling is designed to aid airflow and thereby increase cooling efficiency: if they are removed, cooling suffers. Cowl design and baffle aerodynamics generally aim at producing an area of relative high pressure at the cowl inlets and an area of lower pressure where the cooling air exits. Alterations are not free: they may cost you plenty.

On aircraft whose exhaust systems are visible during preflight, you can look for exhaust leaks, which show up as whitish streaks where the hot exhaust gases deposit various lead compounds and other exhaust products onto the cooler metal. The junction of the engine block and the exhaust pipe is most susceptible to leakage because of differences in expansion rates during heating. Seals are supposed to take up that slight difference, but in the process they may crack. Hot gases may find their way out, and they will toast nearby ignition wires or the electrical harness. After a long period of engine inactivity or in cold weather it is wise to pull the prop through by hand several times to get some oil onto the bare metal surfaces so that a start doesn't have them grinding against one another with no film of oil to lubricate them, and (in cold weather) to loosen up congealed oil so that the starter motor, already straining under reduced voltage in the cold, can get the crank moving a bit easier. Oil does not flow freely when it is cold, so in really cold weather, you should preheat the engine or leave the airplane in a heated hangar overnight.

Aircraft fuel is less volatile than automotive fuels, and it is therefore more difficult to get enough vapor into a cylinder to form an explosive mixture. To compensate for this, aircraft engines have primer systems that inject relatively large doses of raw fuel into the intake manifold. Once enough raw fuel is present, there will be enough vapor for ignition, and, as soon as start-up is accomplished, incoming air will vaporize the remaining fuel. Do not prime a hot engine unless you are absolutely sure that it's called for, with one exception: the vapor lock that sometimes causes difficult hot starts in fuel-injected engines can often be cured by running the electric pump or primer *with the mixture at idle cutoff*. The raw fuel won't get into the engine, but the cooler gasoline that the pump forces into the fuel lines will eventually condense the bubbles of hot vapor that otherwise block the flow of fuel. Leave the pump running for a good minute or so if you decide to try this technique. If you overprime and cause a fire, continue cranking the starter with the mixture at idle cutoff and the throttle fully open. If the engine starts, the airflow through the carburetor will probably suck in the flame, which is usually nothing more than raw fuel burning inside the carburetor. Don't be shy about yelling for an extinguisher early in the game instead of waiting for it to blow itself out. Once it's out, shut down and check for fried wires and smoldering rubber.

The smell of gasoline signals a flooded engine. Russ Elwell of Continental says that his procedure is to move the mixture control to idle cutoff and to open the throttle all the way. Continue cranking until the excess fuel is purged. When the engine catches, which should occur as soon as the air-fuel mix becomes combustible, carefully move the mixture to rich and immediately bring the throttle to idle.

After starting, immediately check for oil pressure. Any engine should show some indication of oil pressure within 30 seconds. In warm weather, you should see a pressure indication a lot sooner; in winter, expect it to take a little longer. If oil was present in the cylinder, it will show up now. Oil in cylinders can foul and short out plugs, causing rough running within the first few seconds. The oil itself won't short the plugs: the fine carbon particles contained in the oil conduct electricity across the plug gap. This is the critical period for the lubricating system of your engine. Between initial start and the conclusion of warmup you must manage the engine so that it is running smoothly (about 1,000 to 1,200 rpm should get the job done) but not producing so much oil pressure that the oil bypasses the engine and takes the shortcut through the relief valves. The oil pump sends oil under pressure to a screen or filter, and there is a bypass valve to allow the oil to flow past them if they are clogged or if the oil is cold and sluggish. The filter bypass is used because dirty oil is better than no oil at all.

Remember that in wet-sump engines the space below the crankshaft collects the circulated oil and lubricates the bottom of the cylinders by splashing, so the proper volume of air above the oil must be maintained. This airspace is downright turbulent, with the crankshaft whirling just inches above the oil. If you have overfilled with oil, you may begin to blow some of it out and maybe even build up enough internal pressure to force the various seals out of their seats, allowing large quantities of oil to leak. On long-distance flights it's okay to fill to the top mark-never higher. Most operators keep the tank a quart or two lower for shorter hops, especially training flights that may involve maneuvers that would slop the oil around. Remember, though, the more oil that is present, the better it will carry off heat. Oil doesn't just lubricate: it cools as well.

Warmup is a good time to perform the mag safety check, which simply involves turning the mag switch to the off position momentarily to make sure that neither mag is hot. Since the function of the mag switch is to shortcircuit the sparking current, it's important to make sure that you'll be safe the next time you pull the prop through by hand. Make the check quickly, though, and at low rpm − just fast enough to make sure that the sparkplugs are indeed dead − for the engine will be pumping an eminently combustible mixture out the exhaust ports all the while, and if enough of it collects in the exhaust system before you turn the mags back on, you'll ignite a spectacular backfire that could at worst blow off the muffler. Warmup rpm settings must be selected with a certain degree of care. You want to keep the engine running smoothly to minimize the strain that roughness can cause. Run too fast and you'll produce high oil pressures that eliminate the flow of clean oil to the innards; rpm settings that are too low may foul plugs and prolong the critical warmup period. Sparkplugs sit squarely in the middle of the burning fuel-air charge. They are designed to conduct away a fixed quantity of this heat of combustion; their ability to do so is measured by their so-called *heat range*. Aviation plugs are fairly "cold;" they conduct a relatively large quantity of heat. At a low rpm they may be so cold that they condense the combustion products of lead additives and scavengers; power settings should therefore be sufficiently high to produce enough turbulence within the cylinder to keep the plug tips well scrubbed.

A lean mix neither hastens warmup nor inhibits fouling; at warmup rpms, which are barely above idle, moving the mixture control will have hardly any effect at all on the actual air-fuel mixture. Fouled plugs often result from an overrich mixture, which produces carbon that eventually bridges the spark gap with a conductive buildup. A leaking primer may well be the cause: if you see black smoke coming from the exhaust during warmup and you know that the air filter is not clogged, suspect the primer and check to see if it is locked.

A common technique to burn off plugs and clear them of deposits is to run the engine up to takeoff power and then to lean the mixture, but a recent Lycoming *Flyer** claims that all deposits on plugs are conductors of electricity to a certain degree and that heating them suddenly will cause the fouling deposits to undergo a chemical change that makes them even better conductors and therefore even more prone to ground out and cause misfiring.

Many pilots place great importance upon having the oil fully warmed before thrusting the power handle forward for the great takeoff strain, but most engine manufacturers say that in temperate climates, the ride from the apron to the active runway will usually prove sufficient to warm the engine for takeoff, even though the oil-pressure reading might still be high. In any case they say that the real index of readiness is whether the engine can take full power application without stumbling. Many pilots try to hasten the warmup in winter by increasing rpm settings or by leaving the cowl flaps closed. The reason for keeping the rpm settings low are manifold, but it is not a good idea to keep the cowl flaps closed in winter to warm the engine faster. Despite the textbooks' admission that at low rpm during the warmup very little air flows through the cowl, it is difficult to find anyone who will bless the closed-flap technique during warmup, even in winter. The only reason that the cowl flaps are present is for climb and for ground engine operation. Even though there may be little airflow from the prop, the propwash will create a little flow by producing some negative pressure behind the cowl flaps. This flow, combined with whatever breeze is available from nosing into the wind, provides enough air to even out the warming of the metal parts − even in winter. The most telling point of all seems to be that keeping the cowl flaps closed doesn't really hasten the warmup, since the oil is relatively slow to warm anyway. Keep the cowl flaps open and in summertime use another trick: align your nose with the propwash of the airplane ahead of you on the taxiway to borrow some cooling wind while you await your clearance.

The best way to manage the warmup is to mentally follow the heat from its origin in the cylinder. The engine will warm from the fire outward: the heat of combustion courses through the conductive metal rather quickly but heats the oil comparatively slowly. Cylinder-head temperatures should therefore be monitored very soon for signs of overheating. After that watch the oil-temperature gauge and remember that it measures the heat of the oil at the coolest point in its travels through the engine. Testing or using carburetor heat during warmup only eliminates the air filter from its position in the airflow to the carburetor and allows dust-laden air to be gulped into the cylinder. Carburetor ice can build up on the way from the ramp to the active, however, so all rules have exceptions: if

---

*Both major engine manufacturers publish excellent newsletters packed with operational rips for their engines. The *Textron Lycoming Flyer* is available from Textron Lycoming Division, Williamsport, Pennsylvania 17701. Teledyne Continental Motors publishes the *Aircraft Power Reporter*, which regularly features a column entitled "Power Notes." Write to TCM's Aircraft Products Division, Post Office Box 90, Mobile, Alabama 36601.

you suspect that ice is present, pull the heat knob, even for takeoff. Be aware of the power loss. Extend the warmup period in cold weather if oil pressure registers above the redline during a power run-up; continue warming until the pressure drops into the green. If oil pressures persistently hover above the red, you may have your mechanic reset the pressure (assuming that you're using the correct viscosity of oil) so that the needle will be in the green during run-up and at cruise.

During power run-up your mag check is intended to reveal three things about the engine: magneto timing, engine condition and power, and ignition-system condition. A difference in rpm drop between the two mags is a sign of bad ignition timing; engine and ignition-system condition are monitored by reading the instruments and using your ears. Nose into the wind and prepare to run up. Fixed-prop engines can be run up whenever the engine accepts the application of throttle without hiccuping on the fuel. Switch to each set of mags in turn, remembering that you are grounding out first one set, then another. Allow enough time on each setting to reveal any defects yet not so long as to foul the plugs (remember − they cool quickly). As each mag is cut out, recall that the accompanying rpm drop occurs because power falls off with only one set of plugs in operation. Rough operation during any portion of the check is usually an indication of a fouled plug or ignition malfunction. Both mags may occasionally drop below the limit stated in the manual. That may indicate the wrong grade of fuel, fouled plugs, improperly gapped plugs, or an incorrectly adjusted mixture.

Should you try a full-power run-up before actual takeoff if you're in doubt about anything? Not according to Joe Diblin, of Lycoming. He says that it is hard on the engine, the prop, and the brakes and that you can accomplish the same thing by watching the instruments and listening closely during the actual takeoff run. Russ Elwell, of Continental, recommends that the propeller rpm not be allowed to drop more than 300 while cycling the props; he also recommends cycling the prop several times on a cold engine to ensure a flow of warm oil between engine and prop. When you exercise the props, you are purging air from the prop oil system and checking the governor. Do not allow the engine to run at high power and low rpm settings during run-up, any more than you would allow such overstressing operation during cruise. Some authorities suggest omitting the prop check altogether.

You're cleared for takeoff. As you push all the handles forward, remember that smooth power application is one key to long engine life. There are dynamic balancing weights on some crankshafts that take different positions at different rpm settings in order to absorb the vibration and to keep the whole crankshaft from resonating to destruction. Abrupt changes in rpm do not allow enough time for the weights to readjust themselves properly to the new range of vibrational frequencies. Rapid application of full power will cause the crankshaft to go out of balance for what may be a critical time.

At takeoff power settings fuel metering, whether by injection or carburetion, is set to provide extra-rich mixtures. The excess fuel absorbs heat and helps cool the cylinders. You may be justified in leaning for takeoffs from high-altitude airports or on a hot day, but you are never justified in using less than full-throttle settings for takeoff. Apply the power smoothly and apply it all the way. Partial power on takeoff causes the engine to lug and creates stresses within cylinders and bearing surfaces. High-altitude takeoffs are accomplished by setting the mixture to the rich side of best power with an EGT gauge or peak rpm during the run-up. With a constant-speed prop but no EGT, lean until smooth operation is obtained. Keep those settings when you take off and watch your engine instruments throughout the roll.

Climb is a period of conflicting needs. If you maintain high rpm and power settings during initial climb, you risk annoying people on the ground. You could climb very steeply at maximum power but with less airflow through the cowl. The alternative is to make a shallower climb with the prop pulled back to a more civilized rpm level. Using a shallower angle also allows higher airspeeds, and, although you may fly a little closer to the rooftops than you would with a max-performance climb, the engine will get more cooling air, the power will produce less heat and noise, and you'll pass through a little faster. Once you are at cruise, it is time to fine-tune the engine to its most efficient and economical setting. Most manuals specify settings of 75%, 65%, and so on. Can you use settings in between? Of course — just approximate the differences between, say, 65% and 75% to get 70%. All prop rpm and mp pairings for the same power produce the same fuel flow, so seek the smoothest setting: reduce vibration and you reduce wear.

Injected engines produce more even distribution of fuel to each cylinder than do carburetors, which force you to set the mix for the leanest cylinder even though other cylinders may run richer than necessary. If you detect a fuel-flow reading that appears inordinately high for the power setting that you're using, it is probably an indication of dirt somewhere in the fuel system, since the fuel-flow gauges are actually fuel-*pressure* indicators, and even a tiny bit of dirt in the injector nozzles or lines creates back pressure.

At cruise power the cylinder head temperature will be cooler at lean mixtures than at best-power settings; at higher power settings the cylinder-head temperatures will rise with a lean mixture. It is important to remember that the engine's designed-in cooling has two components: direct cooling through fin surfaces and indirect cooling through engine oil. At high-power settings an excess of fuel is needed to carry off the heat, which becomes higher than the engine's two cooling components can handle alone. At cruise you can cool the combustion temperatures by either enriching or leaning the mixture. If you enrich, the extra fuel cools the combustion chamber; if you lean, extra air flows into the cylinder; total flow is determined by the volumetric limits of the engine. You can spritz in some fuel, however, which is your only choice. Lean mixtures occasionally produce backfiring, probably because they burn so slowly that they are still combusting when the next charge appears at the intake port. Best-power mixtures differ from best-economy (minimum fuel flow for a given power output) by increasing the working mass of the charge and speeding up the combustion process, which yields more power by converting more fuel to power in a given time.

Carburetor ice can form at any time, but it seems to occur most frequently when the air is below 80 F and the humidity is 50% or better. Fuel vaporization in the carburetor can cause a 70 F drop in the carburetor air temperature and produce ice, even though there is no visible moisture in the air: high humidity is enough. The most dangerous condition seems to be a muggy day with the air temperature about 68 F and the dew point only 2 or 3 F away. If conditions are conducive to icing, you may wish to check for ice prior to takeoff. When you run up and check carb heat, linger awhile with the heat on to see if any symptoms show up. The classic signs of ice are an initial drop in rpm or mp when you apply carb heat, followed by a slow rise. You can expect almost no carb heat during descent − at least not enough to rid your carb of large amounts of ice − for the exhaust system from which the heat is taken cools quickly. What residual heat remains is usually enough to keep ice from forming, so it is better to apply carb heat before the actual descent commences. In cruise the formation of ice is signaled by a slow, steady deterioration of engine performance, both in terms of decreasing rpm or mp and of engine roughness. Apply full heat immediately but don't touch the throttle. Moving the throttle may actually cut off the engine if ice has formed just below the throttle plate. Expect added roughness initially after applying the heat, followed by a smooth rise in rpm or mp. If you do not have a carburetor air-temperature gauge, avoid partial carb heat: all or nothing is best, unless you have the gauge, and understand its use.

There is a difference in the propensity for icing between float-type and pressure-type carbs. Float types must discharge their fuel spray at the point of lowest pressure in the carburetor throat; the venturi effect coupled with the latent heat of vaporization is enough to form ice readily. The temperature drop in pressure-type carburetors is not so great, because they discharge at a pressure closer to atmospheric and therefore create temperatures that are not quite so low. Carburetor heat causes a power drop, because engine power is directly dependent on the density of air and fuel that is packed into the cylinder prior to the compression stroke. It stands to reason that fewer molecules of fuel or air − a charge of low density − will produce less power. Applying carburetor heat effectively lowers the density of the air − that is, there are fewer molecules of air per unit of volume. This thinner air arrives at the carburetor to be mixed with the fuel, but carburetors do their mixing on the basis of weight, not on volume or density. What results is a mixture that is too rich, and you'll have to lean again after applying carb heat. If the outside air is arctic-cold, though, it may have such a high density that the mixture may become too lean. Under those conditions carburetor heat may help restore the mixture to proper proportions. According to Lycoming, in ordinary climes, running on carburetor heat results in an immediate power drop of 3% due to the loss of ram air and the relatively inefficient ducting of alternate air through the heat muff. On top of that 3%, subtract 1% for every 10 F of temperature rise over the standard 59 F at which engines are calibrated for power. The company estimates that the average heat source provides something like 100 F over that standard temperature. Based on that appraisal, carb heat can cost you a total of 13% of rated power.

You're cleared for the descent. You claim that the controller brought you in right over the outer marker and then dumped you? If you have your gear and flaps down early enough, you should still be able to carry enough power on the way down to prevent shock-cooling the heads. Avoid reducing power to the point of letting the prop drive the engine rather than vice versa. Planning ahead for approaches is a habit that those who care deeply about engines cultivate early in their piloting careers. Even with prior planning it is sometimes impossible to work it out so that you can keep the engine warm; when those occasions arise, just set the props full forward to flat pitch and tenderly apply as much throttle (but leave the mixture where it is) as you can without getting into the yellow.

Before shutdown make sure that the cowl flaps are open to allow convection to carry the heat away from the engine. Move the prop to high pitch — if you didn't do it on final — to return all the oil to the engine. In airplanes shutdown is accomplished by shutting off the fuel instead of turning off the ignition, because airplane engines run a bit hotter than other types and have a greater flywheel effect from the propeller; they may diesel if the ignition is shut down before the fuel is turned off.

Beyond these most basic facts about how engines run, there is a great body of detail about each component system within the engine — fuel, cooling, lubrication — each one a study in itself. The FAA's *Airframe and Powerplant Mechanics General Handbook* and *Powerplant Handbook* are satisfying and thorough texts despite their heavy emphasis on large radial engines at the occasional expense of the lightplane engine: most principles apply to both. If reading those does nothing else, it will teach you to talk mechanicese, which is a little like physicianesque, only funnier.

9.

ALL ABOUT OIL

Publisher's Note: While the material in this chapter is not fully current, in that it does not take into account the development of the new synthetic engine oils, none the less it remains a valuable treatment of the subject of engine lubricants.

A piston engine is a mechanical monster struggling to tear itself apart; it is a metal model of neurotic conflicts. Its usually high weight is not required by its eventual power output, as the lightness of very powerful turbine engines shows: the weight is simply there to prevent the engine from destroying itself. Thousands of times each minute bulky parts − many of them made of iron or steel − are subjected to violent accelerations and decelerations and dizzying changes of direction. Surfaces rub and strike against one another at a great rate, while gases burning at more than 1,500 F pummel the pistons, cook the valves, and rattle the exhausts. Add to this the twisting, bending, and prying loads on crankshafts and accessory drives, and a plane becomes a place where ignorant armies clash by night.

It is oil that practically alone saves this suicidal ensemble from itself. It acts not only as a lubricant but also as a coolant, a cleaner, and a cushion between metal parts. The properties of oil are essentially fluidity and a readiness to adhere to solid surfaces. You know how difficult it is to remove oil from a piece of metal or glass. No amount of rubbing will remove that final fine, oily film: the film must be broken down by a solvent before it will disappear. The strong affinity of oil for metal surfaces makes the parts of an engine interact as if they were in fact made of oil: under ideal conditions there would be no metal-to-metal contact in an engine at all − only movement in oil films, the internal friction of which is relatively slight.

The action of oil is sometimes compared to millions of tiny ball bearings. This is misleading: oil molecules do not roll around like ball bearings but slide around like wet noodles, for a typical oil molecule is about 12 times as long as it is wide. Some are branched; some are longer than others. The slipperiness of oil is really just its fluidity. Water is as fluid as oil and generally less viscous (it flows out of a can faster), but it is not usually considered a lubricant because it does not create a durable surface film. Water can be a lubricant, as rain-covered roads show. The point, therefore, is that the essential lubricating property of oil − its "oiliness" − is its ability to form durable films that adhere to solid surfaces.

Under ideal conditions, every moving part of an engine is covered with an oil film: there is no metal-to-metal contact whatever and so there is virtually no wear and very little friction. In practice, however, there is metal-to-metal contact and there is wear, because there is a limit to the film strength of oils: if films of oil are subjected to extremely high pressures, they rupture, leaving metal surfaces in direct contact with each other. High pressures develop in several ways, most commonly because smooth surfaces are not really smooth. The smoothest surfaces in engines are produced by grinding or honing - processes in which the smoothness of the surface is a function of the grit size of the abrasive. In molecular terms, all grits are quite coarse, even the smallest, and all polished surfaces possess a Himalayan roughness on microscopic examination. The oil film envelops the roughness like a gooey ocean, but if two surfaces are rubbed together under high pressure, large irregularities come into collision and breach the oil film, allowing metal-to-metal contact, momentary welding, and wear. The running in of an engine is in fact a period devoted to wearing away large irregularities, accommodating the surfaces to one another.

Another source of wear is due to abrasive material that is carried in the oil. This may consist of airborne dust that finds its way into the crankcase, metal particles freed in the process of wear, or solid combustion products that, along with airborne dust, blow past the piston rings. Any tiny particle carried in the oil film shares with its surface roughness the ability to breach the film and permit wear, with the difference that abrasives suspended in the oil wear by grinding or gouging, while surface irregularities in ground-metal surfaces wear by repeated minute welds and breakages.

The commonest type of bearing in a piston engine is the journal bearing, which consists of a round shaft that rotates in a block, sleeve, or collar: connecting-rod bearings are typical. When such a bearing is subjected to axial loads, as in the power stroke for a given piston, it tends to expel the lubricating film that separates the bearing from the journal. The oil must therefore be continuously replenished under pressure through a system of tunnels and orifices in the crankshaft, connecting rods, and engine block. The pressure is supplied by a pump that collects oil from a reservoir and circulates it through the engine. The pressure maintained is a function of the size of the spaces into which the oil is being forced, the local operating loads, and the viscosity of the oil. The more viscous the oil, the higher the pressure; the pressure drops as bearings wear and clearances increase. Since viscosity decreases as temperature rises, failing oil pressure may signal − besides an instrument malfunction − either excessive temperatures, insufficient oil in the engine, oil with too low a viscosity, or excessive wear.

Fluctuating oil pressure usually signals a low oil level in the crankcase, which is a serious condition that should be corrected immediately: the fluctuations indicate air bubbles that are being fed to bearings. The bubbles produce massive interruptions of lubrication and a sudden sharp rise in wear. Rapidly dropping oil pressure may signal a major failure − an external can-type oil filter that has broken loose, for instance, a leak in the hose to an externally mounted filter, or the catastrophic failure of a bearing. The danger that an air-cooled engine might seize up because of a lubrication failure is not so great as that of a liquid-cooled engine-clearances are larger in the former − but no engine is likely to run for more than a couple of minutes if it is completely drained of oil. Complete loss of oil pressure usually means making a forced landing, though, if the oil temperature remains normal, the problem may be a defective instrument. There are two basic kinds of oil on the market that are approved for aircraft: straight mineral oil and ashless-dispersant, or compounded, oil. Compounded aviation oils are incorrectly called detergent oils — a term properly applied only to automotive-type oils intended for use at comparatively low temperatures. These oils contain metallic-ash additives, which, when burned in the combustion chamber, leave deposits that can cause detonation. Metallic-additive detergent oils are suitable for use in sleeve-valve engines, but, as there are scarcely any sleeve-valve engines in use today, it is almost impossible to find true detergent aviation oil. Because of the ash-residue problem it is not practical to use automotive oils in aircraft engines.

Straight mineral oils are simply distillation products of crude petroleum without special-purpose additives. Distillation of petroleum produces oils of various molecular lengths and configurations, which are blended by the refinery to achieve desired viscosities and viscosity-temperature curves. The general principle is that, for instance, mixing 20- and 40-weight oils in equal proportion yields 30-weight and that mixing an oil whose viscosity remains fairly high at elevated temperatures with one whose viscosity remains fairly low at low temperatures yields an oil with good viscosity characteristics over a wide range. Low viscosity at low temperature is desirable because it reduces the cranking effort for starting and increases the quality of lubrication after start-up; at the other end of the temperature scale resistance to thinning at high temperatures maintains lubrication during climb in hot weather and in other stressful situations.

Straight mineral oil is heavily taxed by the lubrication requirements of an aircraft piston engine; it is inferior to compounded oils, and for this reason it is used in a new engine during the first 25 to 50 hours of operation to permit a relatively high rate of wear. It is impossible to achieve proper break-in with compounded oil: rings will not seat properly, and oil consumption will stay high for the life of the engine. After the break-in period, on the other hand, straight mineral oil should not be used, since it not only has inferior lubricating characteristics but also permits buildups of dirt deposits, which accelerate wear, clog lubrication passages, and increase friction. Its inferior film strength and viscosity-retention characteristics interfere with proper piston-ring sealing and encourage scuffing and high oil consumption.

Compounded oils of the ashless-dispersant type consist of a blend of mineral oils and nonpetroleum additives intended to handle special tasks. The principal problems that affect straight mineral oils are oxidation, acid formation, foaming, loss of viscosity with age, excessive sensitivity of viscosity to temperature, and proneness to leave deposits of lacquer, varnish, sludge, and other undesirables. The deposits are the worst problem; oils compounded to prevent deposits are labeled "ashless-dispersant." If foaming were the worst problem, they would be marketed as "foamless" oils. "Ashless" means that the antisludge additive is not of the metallic-detergent type and will not leave deposits in the combustion chambers. "Dispersant" means that impurities such as soot, carbon, oxidation products, combustion residues blowing past the rings, and other kinds of dirt are held in suspension in the oil-dispersed throughout rather than being allowed to settle onto engine surfaces.

Viscosity characteristics can be improved by adding long-chain molecules — ropy noodles that bind and thicken the whole soup. Foaming is inhibited by silicones that break down the walls of air bubbles. Complex compounds of sulfur and phosphorus interfere with the chemical interaction of oxygen and oil and retard oxidation, while other chemical additives inhibit the catalytic action of metals that normally accelerate oxidation. The dispersant also acts to hold oxidation products — nonoil molecules formed by the action of oxygen on hydrocarbons — in suspension. Antiwear and antiweld agents, such as fatty oils, fatty acids, and phosphorus, chlorine, and sulfur compounds, enhance the ability of oil to lubricate in extremely thin "boundary" films one or two molecules thick, such as those that occur under extreme pressures.

Compounded oil contains its own additives; fresh oil is able to do all that it needs to do, and "miracle" additives such as STP do nothing to improve its action. Oil does wear out in a number of different ways. It suffers loss of viscosity as molecules are sheared and crushed by passage through the valve train and the piston rings: the noodles are broken up and their average length decreases, and consequently their viscosity decreases. Synthetic viscosity improvers are partially effective in preventing such loss. The additives in the oil also become exhausted, just as dishwater that starts with a strong detergent action becomes weak after it has washed a lot of dishes. The dispersants become completely tied up: all the dispersant molecules are busy holding impurities in suspension, and no more are available to handle additional impurities. A good oil filter — and there have been some genuine advances in this area lately — does much to remove particulate impurities, but liquid contaminants held in solution in the oil go right through the filter and turn into corrosive acids through the action of water, which condenses in the crankcase whenever the engine is shut down. Fresh oil contains additives to prevent the formation of corrosives, but, like other additives, they are eventually used up. Throwing in a can of STP on the assumption that it will improve the viscosity of worn oil and replenish additives may do some good — it certainly tends to make everything run more smoothly and quietly for a while — but it does nothing to deal with the problem of acid accumulation. Especially if rings and cylinders are somewhat worn and blow-by and oil consumption are on the rise, there may be a temptation to add STP because it improves ring sealing and reduces oil consumption. It is under precisely these conditions, however, that acid formation is at its worst.

Oil is not an extremely abstruse subject, and there is little to know about it that the engine and lubricant manufacturers don't already know. There are not a great many choices open to the pilot other than to follow closely the recommendations of the engine manufacturers. Experimentation with automotive oils or with "superadditives" may or may not do harm in individual cases, but it is sure to invalidate warranties or even insurance policies if a mishap — even an unrelated one — ensues. The basic rules are as follows.

● Break in the engine with straight mineral oil until oil consumption stabilizes around the recommended value — one quart every 3 to 10 hours, depending on the engine.

● Use *aviation* ashless-dispersant oil at all times thereafter.

● Feel free to mix brands or to add one or two quarts of straight mineral oil to a crankcase full of ashless dispersant if the latter is not available but do *not* add compounded oil or, even worse, STP during break-in.

● Do not fill the engine all the way to the maximum capacity: an oil level between 1 quart below maximum and 1 pint below minimum is satisfactory, but remember that if you take off on a 5-hour flight with the oil a pint below minimum, you may be as much as 1 1/2 quarts low when you land.

● Change oil frequently − more so if the airplane sits idle a good deal, especially in a moist climate.

● Be wary of very low oil temperatures, which may occur in winter if the oil cooler is not baffled off, since a fairly high temperature is necessary to cook away the water that activates acids.

● Check the oil filter or screen periodically for signs of metal particles, which might indicate an incipient failure.

The chief factor in determining the service life of oil is the pattern of engine operation. An engine that is flown frequently − not merely run up − is easier on oil than one that is left idle for long periods. One Shell engineer estimated that an engine that receives daily use and is equipped with a good oil filter and good rings and seals could go 200 hours between oil changes. Frantz, the manufacturer of a new, highly effective oil-filter element, claims 300 hours between oil changes. The engine manufacturers, on the other hand, recommend periods of 25 to 50 hours between oil changes and argue that it is a fairly cheap type of maintenance that can have a dramatic effect on the life and condition of an engine. Much of the uncertainty concerning the correct intervals between oil changes − and, incidentally, the effectiveness of the oil that you are using in preventing wear − can be eliminated through chemical analysis of used oil. The system is usually called spectroanalysis, because part of the testing procedure entails a study of the oil sample by heating, burning, or other methods. Electromagnetic radiation passing from or through the oil passes through a prism and onto a photographic film, where it spreads out rainbow-fashion into many narrow bands arranged in order of wavelength. When different chemical elements are burned, they emit or block radiation of different wavelengths, and by measuring the width of each band of the spectrum and noting its wavelength all the elements present in the oil sample can be detected and their quantity measured. The amount of contamination by solids and liquids is discovered, as well as the presence of worn metallic particles from engine parts. The nature of the contamination usually reveals its origin, and fairly detailed diagnosis of actual or potential engine problem areas is possible. Other tests are sometimes used to assess the condition of the oil itself: how badly worn it is and how its viscosity index − the viscosity-versus-temperature curve − is holding up.

In order for a program of continuous oil analysis to be completely useful, it should start when an engine is new and be carried on with strict regularity, monitoring oil-change intervals and, if appropriate, conditions of operation, so that subtle changes are not obscured in a background of general haphazardness. Such a program can be extremely useful in proper engine maintenance. Even if you have run for a long time without analysis, however, there are certain objective criteria for oil-change intervals, based on the condition of the oil in your engine after a certain amount of use; and the first analysis result will reveal whether you have been running on good or exhausted and contaminated oil. The cost per analysis is on the order of $15, and the handling is done by the FBO. In view of the fact that people have been flying for years *without* oil analysis, it would be silly to say that it should be an indispensable part of every maintenance schedule. For some pilots oil analysis, like seat belts, may never prove useful. There have been cases, however, in which incipient engine damage was discovered by oil analysis, and, if it prevents one engine failure in your entire life, it will be worth its cumulative cost over several uneventful years. If the extra $15 per oil change is not a serious financial burden, oil analysis is worthwhile even if it does nothing more than remove one more area of mystery from aircraft operation.

10.

THE ELECTRICAL SYSTEM

Electricity and plumbing are perennial analogues. The similarity between the way in which water behaves in pipes and the way in which electrons flow through wires may have sparked the expression "juice" for the latter. Whatever its origin, "juice" is like water in a lot of ways — although it is unlike it in even more.

Almost every aircraft owner's manual offers a complete electrical-system schematic as a matter of course, and manufacturers seem to presume that all pilots are capable of reading them. Most can't and, unless they have an atypical remembrance of high-school physics or degree in electrical engineering, quickly flip by that page and read only about what to do in case the lights and the radio go out. The answer to that dilemma usually turns out to be, "Land as soon as possible after turning everything off." That's not good enough. If you are going to take off in any airplane with an electrical system other than the magnetos and ignition harness, you should know what is happening. It is not enough just to say, "Oh well, if it breaks, I'll just do without and go land somewhere." Tell that to the student in a Cessna 150 who wants to raise his electric flaps for a go-around and can't get them out of a 40-degree-down position.

The logical place to begin dissecting the electrical system is with the battery; it is, after all, the only source of electromotive force, or voltage, on an airplane that is parked and tied down. The most common analogy for the battery is the waterstorage tower — the kind that you fly past to find out where you are. Such towers are a source of water pressure because someone has expended energy to pump water up into them. The water will stay there, a potential source of pressure, until someone opens a valve and lets the water run out, at which time the pressure could be turned into work by means of a waterwheel or a similar device. The ordinary lead-acid battery is a source of electrical pressure, called *voltage* or *electromotive force* or just *emf*. You can, if you have to, think of the battery as having been "pumped up" when it was charged; using its chemically-produced electricity to run a motor is similar to letting the water run out of the tower and turn a wheel. You have to recharge the battery after use to restore it; you have to pump water back up the tower to regain the original water pressure.

Aha, you say, but batteries can be run down and will often seem to regain strength for a few minutes' more work, but that is because using the battery alters the concentration of chemicals within it. If some time is allowed for diffusion, fresh chemicals will interface to provide a bit more power. If a water tower were filled with a maze of fine baffles, you'd get the same effect.

To explain why the combination of lead and acid electrolyte produces emf is the stuff of which chemistry courses are made. Let's be friendly about this and just admit the batteries do their job; more important is our responsibility to make sure that the battery is ready to do the job when it's needed. Batteries are an emergency source of power. That the battery provides the source of power for starting airplane engines is largely a matter of convenience. In fact, early aviators, who had no electrical systems, often contrived navigation and cockpit lighting that was no more than flashlights with tinted lenses. Simple dynamos that were windmill-powered by the slipstream (like the spray pumps of many of today's agricultural airplanes) followed; batteries as we know them now in airplanes were not really necessary until the advent of radios and, more importantly, self-starters. Batteries are not complicated. They provide a stated "pressure" of electrons that is ready and eager to accomplish work. The "pressure" of airplane batteries is usually either 12 or 24 volts. Lightplanes use 12-volt systems because they are cheaper than 24; large planes use 24-volt systems not because they are more powerful but because they allow the use of lighter-gauge − therefore, lighter-weight − wire and lower amperage to do the same work. (If you have trouble with that equation, remember that long-distance transmission wires would have to be enormously thick, pipelike things if it weren't for the fact that they use extremely high voltage. Higher voltage means thinner wires − say that 10 times and it's yours for life!)

There are only a few things that can harm a battery. One is heat, either external − the sun's heat in summer, say − or internal − if the battery is discharged so quickly that some of its electrical energy turns into heat. A battery has internal resistance, just like a light bulb or a toaster, and if enough electrons flow through it, it will heat up. This can also work in reverse: if the battery is charged at too high a voltage, it will heat and eventually even begin to boil off the acid electrolyte. A battery that uses a lot of water is almost certainly receiving too high a charging voltage.

If you replenish your battery with distilled water or clean, low-mineral-content tap water and make sure that the battery neither charges nor discharges too fast, it should serve you well. The only way in which it could discharge too fast, incidentally, is with a short circuit: running a number of accessories on battery power for short periods will not discharge a properly installed battery at a destructively rapid rate. It may discharge the cells enough to leave the battery in a weakened condition, however, and allowing it to lie around discharged may produce a condition known as *sulfation*, which is a destructive action on the lead plates by the discharged electrolyte. A discharged battery may also freeze in winter. The most reliable and accurate way to check the charge on a battery is to measure the specific gravity of the electrolyte with a battery hydrometer; the type that is sold for a few dollars at an automotive-supply house is perfect for the job.

Even a fully charged battery, because it depends on chemical reactions that are slowed considerably by cold, will provide less voltage in cold than in warm weather. For this reason experienced pilots are careful to avoid overworking a winter-cold battery. Pulling the engine through before a start helps by loosening up the oil-bound moving parts and reducing the work that the battery must do. A heated hangar helps even more by warming the oil and the battery so that its chemicals are more willing to react and produce electromotive force.

The exact chronology of aviation electrical systems is unclear, but their development obviously hinged on necessity. Navigation lights began to need more power than flashlight batteries could provide; landing lights, a natural product of night flying, were powerful enough to run a small battery down in relatively short order, especially in the cold; windmill-powered dynamos were an interim solution, less sensitive to temperature than a battery and viable as long as the airplane was flying. The engine-driven generator improved matters by providing juice whenever the engine was running. At the same time, adding a generator to aircraft was no simple matter. A battery supplies a "push" of voltage that is always the same, determined − at 12 or 24 volts − by its design and the rules of chemistry. A generator, on the other hand, may sit idle while the engine is turned off, but if it is cranked up to cruise power, that same generator will start putting out voltage that increases directly with rpm. As soon as the voltage climbs (with engine speed) to the point at which it exceeds the voltage of the battery, the generator's "push" will exceed that of the battery, and current will begin to flow into the battery at a completely unregulated rate − probably much too fast for the battery's health. Overvoltage, the scourge of batteries, will have set in to boil away electrolyte-disaster.

To cope with this characteristic of generators, a crude but effective gimmick called a voltage *regulator* was devised. One version works as simply as this: turn the engine up to cruise speed and generator voltage increases; run the current through a wire that is wrapped around an iron core, and as the generator puts out more voltage, an electromagnetic field is produced around the iron core; the electromagnet begins to pull on a metal lever that, in turn, acts to loosen a stack of carbon disks through which the current that energizes the generator's own electromagnetic field courses: the looser the stacks, the greater the resistance to the passage of electricity. Voltage increases; arm pulls on stacks; stacks loosen and allow less electricity to go to the generator field; generator output drops; voltage lessens. In order to produce a slightly greater voltage push than the battery, most voltage regulators keep the generator's output limited to about 14 volts for a 12-volt system: 27.5 volts is typical for 24-volt systems. Modern voltage regulators may use a relay rather than carbon stacks to accomplish this; additionally, though they are called simply voltage regulators, they may also contain mechanisms that limit current (amperage) and prevent the flow of electrons from reversing and running from the battery to the generator. The result of reversed current is the "motorizing" of the generator − that is, forcing the generator to run like a motor on the current from the battery but in the opposite direction of its normal rotation. The generator is the water pump comparable to the battery's water *tower:* turn the generator and you'll get enough "push" to run whatever you need. Because it sends higher voltage to the bus than the battery can (with the voltage regulator preventing overvoltage), once the generator is turning, it pushes out enough to charge the battery and run the airplane's electrical systems.

Alternators are so simple that it's odd that they didn't precede generators. One explanation may be that batteries produce direct current and require direct current to charge them. Generators also produce direct current, but alternators produce alternating current. Now that the transistor age has blessed us with lightweight diodes made of germanium or silicon, alternators can be lighter than generators with the same output, and the alternating current that they produce is easily converted into direct current − a process called *rectifying* − by the diodes. A diode is a device that will pass current in only one direction. By arranging these in a certain way the alternator's output is cleverly turned into direct current. If these diodes are incorporated into the alternator's structure, its output is similar to that of a generator. The main advantage of alternators is that, because of their design, they have an intrinsically higher output than generators and can produce a charging current at engine idle speeds — generators can't. The principal weakness of alternators lies in the diodes, which are even more sensitive to the destructive effects of heat and electricity than are batteries.

One other important difference between alternators and generators is that you can add electrical accessories — the jargon is "load" — onto a generator, and it will obediently produce more and more current until it burns itself out, a noble but self-destructive tendency. If you were to have a short circuit in flight, you could be faced with generator failure for that very reason. Generator systems therefore incorporate a current limiter, a relay that trips the generator off the line when it begins to produce too much for its own health. Fortunately, alternators are self-limited by their design and by the characteristic of the alternating current that they produce. They are rated for a certain number of amperes — 60 amps is a typical figure for lightplanes — and can produce no more. Even if you were to turn on every accessory at once, the total load on your airplane's alternator should never exceed 80% of its rated capacity, unless the airplane is placarded to tell you which accessories must not be used for long periods. Advisory Circular 43-13-1 explains the details of that rule and the exceptions to it.

On airplanes, battery power is engaged by pushing the master switch. The switch is on the panel; the battery is usually mounted far away from the panel. Do the battery wires run directly to the switch? No, because battery cables are large, heavy things — they have to be in order to have low resistance to the battery's meager 12 volts (high voltage, thin wires; low voltage, thick wires) — and because every unnecessary foot of wire means a loss of power to the wire's resistance. Instead the switch closes a contact between two small wires that use just a little battery power to operate an electromagnetic switch called a *solenoid*. If the master-switch solenoid closes, and only if it closes, the battery's *full* power becomes available. If the solenoid should fall, no juice. Many pilots who are confronted by a no-power condition after they've hit the master nod knowingly and say, "Yup, bad battery," when there may well be more to it than that. You should remember that hitting the master is not like turning on the light in the hallway.

Typically, the battery is connected to two things: the bus bar, a kind of electrical outlet into which the airplane's accessories plug, and the starter circuit, which involves another solenoid for the same reasons that the battery needs one. Starter motors require enormous electrical power to turn an engine crankshaft against cylinder compression. Voltage, like water pressure, will do work. But there are only 12 — at most 24 — volts in the airplane, a mere pittance to cope with the huge engine. To make up for low voltage, we cut loose a Niagara-like flood of electrons — high amperage, low voltage and ... thick wires.

The starter switch, wires to the solenoid, cables, and starter motor are normally independent of the rest of the airplane's electrical system. Because of the high amperage in the starter circuit, it is usually wired separately and directly to the battery. Few other accessories enjoy this luxury. In fact, the starter switch may be wired so as to cut off completely the remaining electrical system, both to protect it from the massive jolt of electricity that courses through the starter circuit and to allow the full dose of battery power to go to starting the engine.

Newer airplanes have split master switches. One side of the switch activates the battery solenoid, sending battery power to the bus bar and starter. The other half sends battery current to the alternator field, which is a winding of heavy wires that produces an electromagnetic field within the alternator so that it can produce electricity. What if the alternator fails — due to a bad diode, for example? If you have to run things off the battery for a while, you certainly don't want to waste part of the battery's power on these now-useless field wires that have a very low resistance and that will quickly drain the battery of charge. Instead you can simply cut off the half marked "alternator" or "field" and leave the half marked "battery" on to operate radios and accessories.

One caution about alternators: if you shut one down for some reason while in flight, be sure that you have enough battery power so that, when you hit the alternator half on your split master switch, there is current to the alternator's field windings. Without current to its electromagnetic-field windings the alternator will just keep turning without producing electricity, no matter how many times you recycle the switch.

Only one key element of the airplane's electrical system is left to examine: the ammeter. In airplanes, some ammeters are actually voltmeters that have been calibrated as ammeters. Confusing? Yes, and unnecessarily so. Airplanes have two kinds of ammeter indication (the "idiot light" appeared briefly and unforgivably in airplanes but seems to be disappearing): one is the ammeter with its needle set on a center index mark, with a way of indicating battery charge or discharge to either side; the other type, found most notably in Piper aircraft, has a needle that indexes on a zero mark on the left end of the scale, then moves upward as it measures the load on the alternator. What's the difference between the two? The meters are really the same but are calibrated differently. The principal difference lies in the way that each is hooked up. The center-zero type is hooked up between the *battery* and the bus bar. It tells you only whether current is flowing into the battery or out of it and nothing about how much the alternator is putting out nor how much voltage the system is running on. The Piper kind is hooked up between the *alternator* and the bus bar: if it indicates zero while you're flying, you know that your alternator has failed; by moving upward on the scale it tells you how much current, or load, the alternator is meeting a demand for. The trouble with both types, however, is that they tell you nothing about the voltage being supplied to the bus and the battery. Voltage regulators are marvelously reliable — until they fall. Large airplanes usually have some way to measure voltage, commonly a switch on the same meter (voltmeters and ammeters are essentially the same, differing only in how they are wired to a calibrated resistance), and it is high time that singles also have them. Overvoltage, after all, will do more than just boil battery electrolyte away: it will also ruin radios, which are very sensitive to fluctuations in voltage. That is why you should turn avionics systems off when you use the starter circuit: to protect them from voltage "spikes" — momentary, transient peaks of voltage that are higher than the radio's innards can cope with. Of course, a voltmeter will also warn of a low-voltage alternator before the battery is bled of charge. Isn't one little voltmeter worth the trouble to avoid the inconvenience of having no starting power at some remote grass strip?

What should you do if something falls? If the source of emf weakens, be it battery or alternator, the answer is, disappointingly, to turn everything off until you can locate the trouble. In flight, you may read that to mean turning off everything that's unnecessary. In practice, it means turning off *everything* so that you'll have some power left to land with, especially at night. It's not so bad as it sounds: you can make quick checks to ensure that you're on course by flicking battery power on, checking omni indicators, and flicking if off again. Save the battery for the sustained power demand of an instrument approach, if that's coming up. The specific procedures for troubleshooting an electrical malfunction vary with different makes, but owner's manuals give ample instructions on how to recycle things if breakers trip and fuses fuse. A remarkable number of pilots forget that turning off the master doesn't bother the engine a bit. Airplane spark plugs connect to magnetos, which are totally independent of the remainder of the electrical system. As long as the mag switch is on, the engine will run until the fuel runs out. The only thing to be concerned about is that the engine instruments depart with the rest of the electrical instruments when you shut down the master. Occasional use of battery power will allow you to scan them when you check your course.

Experienced electricians claim that most problems are due to loose wiring, broken insulation, or some such niggling and elusive flaw rather than to the actual failure of a component. The aircraft electrician who carefully probes the wiring harness with a meter is the man to choose over the fellow who listens to your symptoms and then says, like the pilot who can't start, "Yup, bad battery." So simple a defect as a loose connector or a corroded battery terminal can raise circuit resistance to the point where voltage drops to the nonoperational range. Intermittent operation is one of the harbingers of that kind of harness failure and should be mentioned to the service man in any description of symptoms. If his eyes roll heavenward, that's only because he knows that he has a hunt on his hands before he finds the bug. A simple rule of thumb is that, if a breaker or fuse opens after being set a third time, leave it open and seek help. If fuses pop in flight, there is nothing wrong with grabbing other nonessential fuses (as long as the fuse rating is the same) to restore operation when you need it in a hurry.

This discussion has moved from the battery to the bus bar and left you to figure it out from there. There are too few common denominators among the systems that plug into the bus to discuss them in one basic treatment. Read texts, read owner's manuals, and ask electricians, but *do* try to know what's happening when the juice flows.

11.

TURBOCHARGING

As any E6B owner can tell you, the implications of indicated airspeed become increasingly grandiose with increasing altitude. If we indicate 150 knots at sea level, it means that we are going about 150 knots; at 5,000 feet, it means 162 knots; at 10,000, 174; 189 at 15,000, and 208 at 20,000. If we made it up to 30,000 feet, we would be truing out at a respectable 245 knots. Even supersonic airplanes generally achieve their remarkable speeds only at high altitudes. Flying at Mach 3 (over 1,700 knots) at 70,000 feet requires an IAS of only about 515 knots. The reason is − or at least should be − well known: the density of the air in the earth's atmosphere decreases with increasing height above the surface, and airspeed is ascertained by measuring the force of impact of the air striking the inlet of the pilot tube. Impact force is a function of velocity and mass, and, as the density and therefore the mass of a given volume of air decreases with increasing altitude, the impact velocity must increase in order to give the same pressure reading on the airspeed indicator.

Not only the pitot inlet but the entire plane is subject to the density of the air through which it is moving. Air may not seem like much to move through, but at sea level, the air in a Bonanza − sized tunnel a mile long weighs about 10,000 pounds, and trying to get it out of the way at 170 knots is much harder than it is at a walk.

Just as the decreasing density at higher altitudes reduces impact pressure on the pitot and therefore reduces indicated airspeed for a given true speed, so it also reduces drag. It follows that, with a given amount of power available, the airplane can thus achieve an ever higher true speed as it gains altitude, or, to put it another way, with a steady power supply, indicated airspeed can be kept the same at high altitudes as at low.

The key to taking advantage of increased speed due to low air density is to have constant power available. Since engines breathe air just like people, they are similarly subject to their own kind of hypoxia. They burn gasoline and release energy in a fixed proportion to the air that they breathe: 1 pound of gasoline to about 15 pounds of air. The engine breathes by moving a piston down in a cylinder and opening a valve to allow the void thus created to be filled with outside air. At high altitude, the weight of a cylinder-full of outside air is smaller, and the weight of gasoline that the engine can properly burn is also reduced so that it is able to produce proportionately less power. The solution is to provide the engine with the equivalent of sea-level air − air that is compressed to sea-level pressure or density. The compressing is done by a centrifugal pump driven by a turbine propelled by exhaust gases. Since the energy of the escaping exhaust is normally wasted, driving the turbocharger puts no additional load on the engine except for a slight increase in exhaust back pressure.

The turbocharger unit is small but tough. The turbine and compressor wheels are each about 3" in diameter, and the shaft joining them is not much larger than a pencil. The pair spins at about 100,000 rpm, with the turbine immersed in exhaust gases at 1,500 F. The turbocharger operation is modulated by a waste gate − a controller relief valve in the pipe that carries exhaust to the turbine. If little or no boost is needed, the waste gate is open, and exhaust gas escapes through it rather than through the turbine. As the waste gate is gradually closed, more and more gas is forced to go out through the turbine, which spins at an increasingly high speed, producing more and more boost until its full output is reached with the waste gate closed.

The maximum compression available from the turbocharger is limited by the heating that takes place in the compressor-discharge air. Air pumped through the compressor is heated as it gets denser. At a compression ratio of 2.2 times ambient, the discharge air emerges at around 200 F. Above this temperature, detonation may be triggered in the engine. Assuming that the maximum discharge pressure that the engine will accept is sea-level pressure, the altitude at which that pressure is 2.2 times ambient is 16,000 feet, which is the maximum altitude at which the engine will develop 100% of rated power. It is called the *critical altitude*. (The critical altitude could be increased by placing an intercooler between the compressor and the engine to lower the temperature of the incoming air while maintaining its pressure, but the advantages of having full power available above 16,000 feet are not great enough in most light aircraft to justify the added cost, weight, and complexity of an intercooler.)

Above the critical altitude the waste gate is fully closed, and the discharge pressure, or "upper-deck pressure," ceases to depend on the waste-gate setting and begins to depend on the exhaust-gas output of the engine − in other words, on the power setting. The turbo is now said to be *bootstrapping*. This means that, unless power changes are made very gradually, the turbo and the engine start chasing one another around. Increasing power increases exhaust output, which increases turbine speed and upper-deck pressure, which increases manifold pressure, which increases power, which increases exhaust flow. Pilots sometimes complain of manifold-pressure drift and of an inability to maintain a stable power setting with turbos. The problem can be minimized by moving the throttle (or turbo throttle) slowly, especially while in the bootstrap mode.

One of the two types of turbochargers available for light aircraft − the Rajay − controls the waste gate by means of a second throttle. This vernier control is used after the engine throttle has reached its full forward position, usually at around 6,000 feet. The second throttle is used to keep manifold pressure constant by closing the waste gate and introducing as much boost as is required by the ambient conditions. Most of the turbocharging systems available as options from aircraft manufacturers are built by Garrett AiResearch and use a hydraulic automatic-feedback control to operate the waste gate. The engine controls are the same as in a normally aspirated installation. Upper-deck pressure − the pressure upstream of the throttle − is usually compared with a reference pressure generated by a spring, and the waste gate is closed as much as is needed to keep the throttle at sea level. The pilot then sets his power in the normal manner, just as though he were flying at sea level. The less manifold pressure he calls for, the less drain is placed on the air supply, and the lower the turbocharger output needs to be in order to maintain sea-level upper-deck pressure. For example, if, after climbing to altitude at 25/2,500 the pilot throttles back to 23/2,400, the deck air will "feel" the reduction in manifold pressure as an increase in deck pressure (caused by a change in the position of the throttle butterfly). This causes a valve that meters engine oil pressure to the waste-gate controller to open slightly, reducing the amount of exhaust gas going to the turbocharger and thus reducing the deck pressure as well. If it is properly adjusted, the system will always balance out at sea-level pressure (a deck pressure of about 30" Hg).

A further refinement in controllers is found in the system used in the Turbo Navajo, which supplies upper-deck air at sea-level density rather then pressure. Here the trick is to replace the spring reference with a barometer-type reference that compares deck pressure plus temperature with a trapped volume of dry nitrogen that is immersed in the deck air and shares its temperature changes. This system is also designed to provide sea-level boost − that is, it supplies an overpressure even at sea level in order to increase the maximum engine output. Maximum manifold pressure on the Navajo's engines is 40", which lets the engines − normally rated at 290 hp − put out 310 for takeoff. Although the Navajo's turbocharging system provides some power advantages even at sea level, the principal benefits of turbocharging are generally not felt until one spends a good deal of time flying into and out of airports with very high density altitudes or while cruising for long periods at altitudes above 15,000 feet. Rarely, however, do all the advantages of high-altitude flight combine to produce a dream flight. *Flying*'s former Assistant Editor Paul Garrison once flew nonstop from Portland, Oregon, to Harrisburg, Pennsylvania in a Cessna Turbo-System Centurion, showing a true airspeed of over 220 knots and a groundspeed, with the help of a particularly fortuitous tailwind, of over 350 knots. The trip took about six hours. This is the kind of flying that turbocharger manufacturers would like you to believe befalls every turbo owner. In fact, the practical realities of turbocharger operation are somewhat less dazzling.

The actual speed advantage of the turbo is real, but since it appears mainly at high altitudes, time to climb must be taken into account, and block speeds turn out to be only sightly better than with a normally aspirated engine. The advantage is usually on the order of 10% or less. Even if the advantage were as great as 10% on every flight, you could use it only on flights on which the wind was favorable and even then only on flights long enough to justify the climb to altitude − flights of three or four hours at least. Assuming, however, that you manage to wrest a continuous 5% speed advantage from your turbo, averaged over all your flying time, it is certain that that advantage would not translate itself into a reduction in per-mile operating costs. Though the turbo itself is a simple apparatus, it still requires some service, and its controller is as liable to break or go out of adjustment as is any other precision device (such as a constant-speed prop, which it resembles). TBOs on turbocharged engines are shorter than on normally aspirated ones, and overhaul costs are higher. For example, the 285-hp Continental IO-520 has a 1,500-hour TBO, and overhaul costs average around $2,600. Turbocharged (as in Turbo-System Centurions), the same engine drops 100 hours on the TBO and rises between $700 and $1,000 on the overhaul. The turbocharged IO-320s in Turbo Twin Comanches have about the same overhaul cost as normally aspirated engines, but a TBO of 1,200 hours as opposed to 2,000 for the plain engine. Even minor routine servicing may be complicated by the encumbrance of the turbocharger under an already-crowded cowling.

Another major problem is that, until cabin pressurization — supplied by the same compressor that supercharges the engine — becomes prevalent, using the turbo means using bottled oxygen. Add the cost and inconveniences together and it develops that the popular image of the turbo as a little bundle of free speed is a misconception. Speed is perhaps the turbo's most attractive and easily appreciated selling point, but it is not the most *practical* aspect of the device. The practical aspect is the takeoff and climb performance for aircraft used in high-altitude, heavy-load work and the enhanced single-engine performance of twins. It is in air-taxi, sightseeing, and cargo-hauling operations in mountainous country that the turbo comes into its own as an important safety factor and a sound investment. If that is true of single-engine airplanes, it is doubly so of twins. Most light twins have single-engine ceilings of 6,000 or 7,000 feet. Many a summer has never seen a standard day, however, and density altitude is a problem that rides with all twin-engine operations. Turbos solve that problem.

Turbochargers have a logical place in light-aircraft engines. There is no reason to sit idly by and watch power dribble away as you climb up to altitudes that on other terms should be the most useful for cruising flight. It is likely that a system for maintaining sea-level upper-deck pressure up to the floor of the positive-control airspace will eventually be a standard part of most engines (assuming that the piston engine survives emission controls, which it may not be able to do). Equipped with a bleed system to provide cabin pressurization, a well-integrated turbocharger would bring the conventional piston engine as far as it is likely to come in its effort to resemble a turbine.

If the day comes when most light-aircraft engines are turbocharged and many cabins are pressurized, pilots will face a new set of requirements for basic proficiency. Although the turbocharger is said to get you up and over the weather, it may also take you up and down through more weather or up into worse weather, with greater exposure to icing and a longer way to go to get away from it. More sophisticated flight planning would be required than is customary today, and most flights would customarily file IFR in positive control airspace. The gap between a low-altitude trainer and a high-altitude traveler — even one the size of, say, a Cherokee Arrow — would widen. Though the turbo itself is simple, it opens the door to complexities beyond the realm of the merely mechanical. Think twice before going to turbocharging: you might be biting off more than you can use.

12.

PRESSURIZATION

Say you're at 7,000 and holding for a clearance into La Guardia. Nagging little delays have compounded all day, and it's nearly seven o'clock; the dinner reservations went down the drain an hour ago, and you're wondering whether you'll even use the theater tickets. In a glow of pride in your new twin you invited two friends to fly to the city for a night of dining and theater, and they, flattered and excited at the prospect, accepted. Now you're wondering how to rescue a sure social disaster when the controller calls. You're cleared for the approach into La Guardia. With sudden relief you push the nose over and then realize where you are. You've got 7,000 feet to lose in the four minutes between holding and touchdown. You explain lamely that you'll be losing altitude rather quickly, that there's no cause for worry. Both guests are pale, wide-eyed. Their ears know that they're diving, thank you. Down you plunge, the airplane shrieking like a Stuka. As you're locking up after parking, both guests try to cover their shakiness with thin smiles. She has a slight cold and hasn't yet cleared her ears. He makes some crack like, "I think I'll stick to the airlines." You wish that you'd had a pressurized airplane during that descent, but they're expensive and you'd need a graduate course to understand the system. Anyway, only big airplanes are pressurized, right?

When Cessna announced the pressurized 337 Skymaster, it was a clear signal that pressurization was moving down the general aviation price ladder. Even pressurized single-engine airplanes are not the fantasies they once were: the first ones may break into the market in three years. If that smacks of overweening optimism, it's probably because until now pressurization and its associated systems have been thought to be too costly and complicated for lightplanes. Although the pressurized cabins introduced on airliners and military aircraft were complex, unduly complicated, and overdesigned affairs, time has eroded the machinery so that pressurization systems now seem absurdly simple once they are dissected into their component parts.

All of them start with a source of pressurized air. The source can be a pump such as the hydraulically driven centrifugal compressor on the Aero Commander 720 Alti-Cruiser, one of the first general aviation airplanes to have a pressure cabin. Each of the Commander 720's engines had a hydraulic pump that supplied the drive for the compressor. It was a Rube Goldberg setup at best, and the airplane never sold well. The more likely source of pressure in contemporary airplanes is bleed air from the turbocharger in a piston engine or from the compressor stages in a turbine engine. Using bleed air is like diverting a small creek off the enormous river of compressed air in the turbocharger or compressor stages. You borrow a few pounds of air from that vast reservoir, and it's never really missed. The air, now under pressure, usually passes next through a sonic venturi − a restriction in the air line to regulate the flow and limit the pressure of incoming air.

The air has now progressed along the system to the point where there's nothing between it and the cabin except coolers and heaters to keep the temperature of the incoming air at livable levels. To put the pressurized air into the cabin is one thing; keeping it from leaking out is another. For this reason pressure cabins are sealed and beefed up to take the extra force of the air trying to push its way out.

Sealing a cabin is a matter of attention to detail: you have to find all the nooks, crannies, and cracks where the air could find its way out and then seal them up. If you can rid a house of cockroaches, you have a bright future in pressure-cabin design. Doors and other openings get extra latches and seals, with the seals usually being of the bladder type that expands and conforms to the space between the door and the doorframe in order to make a tight fit. Reinforcing the airframe is a more serious business. Four or five pounds per square inch of air pressure doesn't sound like much; what counts, though, is the surface area inside the cabin. Five pounds per square inch begins to add up to tons when you start multiplying by square inches of cabin surface area.

All the extra strength is designed to cope with a push from the inside out, never the reverse. A pressure cabin is somewhat akin to a balloon with tape around it. The balloon may be prevented from expanding and bursting by the tape, but nothing keeps it from collapsing when all the air is let out. Fortunately, airplanes fly up, from higher to lower pressures, and are inflated − like the balloon − from the inside. If pressurized airplanes should ever descend to where the pressure outside is greater than that inside, there's a one-way valve to let in the outside air and keep from collapsing as the balloon did.

If all this talk about balloons makes pressurization sound like kindergarten stuff, it isn't. When the first de Havilland Comet airliners underwent sudden catastrophic airframe failure at altitude, engineers were at first stumped about the cause. After a massive job of aeronautical detective work, the source of the problem was finally traced to metal fatigue that was directly attributed to pressurization strains on the airframe. As the air came in and out, the cabin and airframe had to expand and contract slightly with the load each time. Those minute movements began to add up until fatigue caused a part to crack and fail. Pressurized cabins must be fatigue-tested thousands of times by cycling the air in and out, in and out until all potential sources of structural failure are isolated and strengthened or redesigned. Cessna's 337 pressure-cabin capsule was cycled 42,000 times on a hydraulic test rig. That's the equivalent of two flights a day, seven days a week, for more than 57 years.

The object of pressurizing a cabin is to pump enough air into it to keep it at sea-level pressure. The pumped-in air is called differential pressure and is measured in pounds per square inch. The more you pump up the cabin, the higher you can fly and still maintain a cabin with sea-level air pressure. High differential pressure is a favorite selling point for manufacturers and getting it sounds simple enough, but, like buying on credit, you can go just so far before life catches up with you. The limiting factor in the case of most airplanes is the

61

strength of the airframe. With a weak airframe, you may be able to pump only one or two psi into the cabin. With sea-level pressure at 14.7 psi on a standard day, once you climb such an airplane to the altitude at which the pressure outside is 2 pounds less than inside, the cabin will have reached its maximum differential pressure and it will have to start climbing with the airplane. If you build your airframe stronger, though, and design a system capable of pumping 14.7 pounds per square inch of air into it, you'll have a spacecraft. Passengers in such an airplane would never experience even the slightest change of pressure; the cabin could be held at sea-level pressure no matter how high you flew it.

The manufacturers' problem is one of compromise: they need a pressure capsule strong enough to withstand the highest possible differential pressure and thereby enable high cabin pressures at altitude. Adding strength means adding weight, and putting weight into the airframe will eventually detract from performance if it isn't complemented by a power increase. The ideal is to build an airplane that maximizes all these factors in about equal balance: speed, ceiling, useful load and differential cabin pressure. If you fall too low on the first three factors, the result is obvious; if you fall too low on cabin pressure, however, all the work and money spent to make a pressure cabin in the first place may be sacrificed. Nobody loves a pressurized airplane that maintains only a couple of thousand feet between cabin altitude and actual altitude.

So far, we have air coming in under pressure and a cabin that's reinforced and airtight. Our next problem is to control the pressure. The way in which most pressure systems do that is by controlling how fast the air leaks out of the cabin. The "leak" is a valve whose only job is to vary the flow of air leaving the cabin. The more the valve opens, the faster air will flow out of the cabin; pressure in the cabin will drop, and the cabin will feel as if it's climbing. If the valve closes, pressure will increase in the cabin, and the cabin will feel as if it is descending.

In case you're wondering, it is possible to descend a pressurized cabin below sea level. In fact, that is exactly what happened on one of the first recorded attempts at pressurized flight, in 1921. The pilot on that flight, Lt. Harold R. Harris, initially had some trouble locking the door after ascending to 3,000 feet. The designers, pessimists all, had doubled their air-compressor capacity because they were worried about losing too much air through the openings around the control cables. Soon after he got the door closed, though, Harris' cabin altimeter showed him at 3,000 feet *below* sea level, although the airplane's actual height was some 3,000 feet above sea level. The flight would have gone well except that the temperature inside the pressure tank had climbed to 150 F. Harris had no way to control the cabin pressure, so he landed. The heat problem on Harris' flight stemmed from a basic law of physics: when air is compressed, it gets hotter. Some airplanes even use the heat of pressurization to keep the cabin warm, but if it's summer and plenty warm already, cabin air must be cooled after it's compressed. At high altitudes air temperatures are low enough for passengers to be comfortable without air conditioning. Down lower pressurization without air conditioning means virtual suffocation, and pressurized airplanes without air conditioning are an unhappy compromise. Basically, then, that's the complete system: a source of pressure, an air conditioner and/or heater, a pressure container for passengers and crew, and a controllable outlet.

Ways of controlling pressurized cabins vary — some are vacuum-operated, others electrically powered — but most controllers are simple. The simplest system of all has no real control, just an on-off switch. At an altitude set by the pilot, the system cuts in and pressurizes the cabin. While the airplane climbs, the pressure system holds the cabin at the cut-in altitude until the differential pressure reaches its maximum. As soon as that point is reached, the cabin climbs with the airplane. What limits the differential pressure is the strength of the airframe, not the pump. The pump could continue pumping more air in and keep on pumping beyond the maximum differential pressure, even if it took a while to do it. A pressure-sensing valve determines what max differential pressure has been reached and automatically begins to let air out of the cabin so that, although the pump keeps pumping, air is leaked out at a rate sufficient to ensure that the pressure difference between the outside and the inside of the cabin never exceed the maximum set for the airframe. Most military aircraft and some general aviation aircraft use this simple system. Airliners and an increasing number of general aviation airplanes use another slightly more expensive and sophisticated system, one that enables the pilot to select pressurization from sea level up to altitude. There are two controls in this system: one knob sets the cabin altitude; the other sets the rate of cabin-pressure changes during a climb or descent.

Here's how a typical flight with this system might go: Taking off from Newark, New Jersey (field elevation 18 feet), climb the airplane to 20,000 feet for a trip to Wilmington, Vermont (field elevation 1,956 feet). Let's assume that it takes 20 minutes for the airplane to reach 20,000 feet. When we reach that altitude, our max differential pressure − 4.2 pounds, say, to choose a figure typical of lightplane pressure systems − will give a cabin altitude of 8,000 feet. There is one simple choice at this point. We could set the cabin-altitude control at sea level and wait for the inevitable point where max differential pressure would be reached, at which point the system would then automatically leak air and the cabin would climb at 1,000 fpm with the airplane. But we have 20 minutes before the airplane will reach cruise altitude. Why not use the rate-of-change control to give our passengers the comfort of a more gradual cabin climb?

The highest cabin altitude we expect to reach for the flight will be 8,000 feet, so we can set that number on the cabin-altitude controller right after takeoff. We have 20 minutes before we'll reach cruise altitude, so divide 8,000 by 20 to find the rate of change we can expect. The answer is 400 fpm, a rate that is quite comfortable on the ears. Set the rate control for a 400-fpm change and climb. While the airplane is traveling from sea level to 20,000 feet, pressure air to the cabin will be balanced by outflow to provide a smooth, steady, and gradual rate of change in the cabin, a rate that will never exceed a comfortable 400 fpm.

Now, we're at 20,000 feet with an 8,000-foot cabin, and it's time to set up the descent. Destination elevation is about 2,000 feet, so there's some 18,000 feet of airplane altitude to be lost at, say, 1,500 fpm. A descent at that rate should take us 12 minutes. We'll have 12 minutes to descend the cabin as well, but this time we only have to descend it to 2,000 feet, to match destination elevation. Plug that number into the cabin-altitude controller and, since we have 12 minutes to lose 6,000 feet of cabin altitude, dial in a 500-fpm rate of cabin descent. We can treat your passengers to a gradual cabin descent while the airplane plummets down at three times the cabin's rate.

What if the pilot had been careless and set the cabin altitude at sea level for a landing at Wilmington? He would have landed with considerable pressure inside the cabin, probably enough to make it difficult to operate the door latches. If the door had opened, there would have been a popping noise as pressure was released, but nothing dangerous. Instead of letting that happen, though, there is a device known as a dump valve to get rid of any excess pressure. On landing, the weight of the airplane on the gear activates a switch and pressure is equalized between inside and outside.

The pilot could have made the same mistake in reverse and set the elevation at Wilmington (2,000 feet) into the controller for a landing at Newark. As soon as the airplane passed through 2,000 feet on its descent to Newark, a simple one-way valve would have allowed outside air to come in and balance the cabin air, thereby preventing the airplane from landing with less pressure inside than outside. Remember that pressure cabins are weak when it comes to dealing with negative differential pressure.

If it hasn't become obvious yet, the primary excuse for springing the extra bucks for pressurization is not high-altitude capability but comfort. Even if you never take your airplane above 10,000 feet, you can still use a pressure cabin. Steep descents can become a way of life and without guilty excuses.

Pressurization is not without certain hazards, the most widely advertised of these is sudden decompression. Decompression conjures up memories of airliners fractured like eggshells, human guinea pigs in test chambers with beakers of water boiling in the near vacuum, and Keir Dullea's "naked" walk in space in the movie *2001: A Space Odyssey*. The threat of a hijacker's stray bullet is more of a danger to fuel and control lines than it is to the thin skin of an airliner, but if a large transport cabin is seriously ruptured, the wind created within the cabin as the air gets sucked out will carry with it anything loose. Its force is even great enough to throw people around. Another danger at very high altitudes is that a pilot could lose consciousness within 15 seconds after loss of pressure. Airliners have oxygen masks that automatically drop down in front of passengers in the event of decompression; if decompression occurs, the pilot must make a quick switch to oxygen and fly the airplane into a quick descent. For this reason regulations specify that oxygen must be kept close at hand in the cockpit. With the exception of high-performance business jets, few general aviation aircraft fly as high as the airliners and the maximum differential pressures are lower for small airplanes, making the effect of decompression somewhat less drastic.

Pressurization controls, systems, and cabins have progressed to the point where the associated hazards have become acceptable risks when balanced against the comfort that they provide. At lower altitudes, where the effect of decompression is reduced, pressure cabins even provide an extra safety factor. Although the regulations say that 12,500 feet is the critical altitude at which pilots need extra oxygen to function, it is widely conceded that factors such as cigarette smoking may lower the critical altitude for some individuals. By providing a low-altitude environment pressure cabins provide an extra margin so that pilots whose physiology falls into the twilight zone won't have to take chances with their reflexes and judgment.

After years of being restricted to large corporate hangars, pressurized cabins suddenly seem about to become popular. Once billed as a necessity for high-altitude flight ("Now you can get above the weather..."), pressure cabins are at long last being pushed for what they really are — comfort options. The additional cost no longer seems out of reach in proportion to the total cost of the airplane. Cessna's pressurized 337, with a modest 3.35-psi maximum differential pressure, was not equipped with a pressure cabin so that the airplane could strain toward the stratosphere: it is simply a more comfortable airplane with the addition of pressure. If Cessna's P337 can make a sales success out of that approach, other lightplanes are sure to follow.

II. THE COCKPIT

The instrument panel of an airplane is like television: what it says is obvious (sometimes), but few pilots know how it manages to say what it says. Most people find electricity difficult and electronics worse. After a look at some of the finer points of how instruments communicate with people, the discussion turns to the difficult world of avionics. The last seven chapters are unusually long for this collection because the subjects are broad and diffuse. They avoid explanations of how VOR and ADF are used to navigate − best explained in the cockpit − and try to get at how they work and why they sometimes fall. For radios, there is at least the everyday experience of radio communication to start with; for autopilots and flight directors there is nothing in everyday experience, and the discussion dives deeply into them to surface with the fundamentals on a level no more rarefied than that of lawn mowers and alarm clocks.

13.

UNDERSTANDING ENGINE INSTRUMENTS

Aside from the principal engine instruments, which register rpm and manifold pressure, there are several others whose small size and frequently inconspicuous position make one prone to ignore them. They include the oil-pressure, oil-temperature, fuel-pressure, and cylinder-head-temperature gauges. All four commonly give the same monotonous indications from day to day so that one barely notices them − until, often too late, one or another of them has changed its habits.

All these instruments are characterized in most modern panels by a green operating range with redlines at either end. In general, any steady indication in the green is a sign of good functioning; any unexpected fluctuation is a sign of possible trouble, which may be in the instrument rather than in the engine: if one gauge acts up while everything else reads, *sounds*, and *feels* normal, then the gauge is suspect. Otherwise your best bet is to assume that the gauge is telling the truth and to act accordingly. Though the basic message of an unusual instrument indication is usually "land," it may be possible, with a sufficient understanding of the engine and how its instrumentation works, to judge how urgent the advice is, what the reason for it is, and what action should be taken on the ground to set things right again.

The primary function of the oil-pressure gauge is to inform the pilot that oil pressure exists and that it is within specified limits. The oil pressure is produced by a pump that takes oil from the sump and sends it through a series of passages, called *galleries*, to the various bearings, sleeves, and bushings that must remain drenched in oil in order to operate properly. Air-cooled engines depend heavily upon their lubricating oil for cooling: heat picked up by the oil on the way through the engine is dissipated either through the sump walls, in small engines, or in an oil radiator. The flow of oil through the oil radiator is sometimes controlled by a thermostat, which operates in the same manner as the thermostat in an automobile radiator: when the coolant (the oil, in this case) is at too low a temperature, the thermostat retards the flow through the radiator, reducing heat loss and raising the temperature of the oil. The system is completed by a pressure-relief valve at the pump end; its purpose is to prevent a pump overload in case of a high resistance in the lubricating system − caused by closing the radiator thermostat, for instance.

Oil pressure may be read − "picked off" − either just after the pump, in which case total oil pressure is reported, or just before the sump, in which case residual pressure is read. A typical pickoff at the pump end reads a consistent high pressure and gives a nearly instantaneous indication of pressure upon engine starting; for most purposes, however, it is less informative than a pickoff midway in the system or at its end. If the pickoff is located near the end of the lubricating system, as on Continental O-300 and O-470 series engines, the gauge indication lags on start-up, but when the pressure comes up, you know that oil has made its way through the entire system and that the engine is receiving lubrication. A pickoff at the pump end might be misleading in some cases: for instance, on a very cold engine congealed oil in the galleries might cause quite a bit of oil to pass through the pressure-relief valve, giving a good pressure indication but no lubrication. The pressure read from a pickoff at the sump end is actually "bonus" pressure − the pressure remaining after most of the engine has received lubrication. If the pressure indicated at normal operating temperature at idle speed is within specified limits, the end pickoff also testifies to the good health of the engine bearings.

Even without knowing the type of pickoff involved, however, the pilot can still learn a lot from his oil-pressure gauge merely by comparing readings from day to day. Sudden abnormally low oil pressure can indicate low oil quantity; this condition is usually accompanied by an abnormal rise in oil temperature. The same symptom might have other meanings, however: oil pressure will be low but consistently so if oil of too low a viscosity is used (30-weight rather than 40- or 50- in warm weather, for instance). A gradual loss of oil pressure over a period of time might indicate clogged filters and screens − a condition that usually arises from the operator's failure to change oil at the prescribed intervals. If engines equipped with thermostatically controlled oil radiators are improperly winterized for extreme cold, they may experience a loss of pressure and a rise in oil temperature shortly after takeoff. If flight is continued, both gauges may eventually give redline indications. The cause is oil congealed in the radiator, which fails to clear out when the thermostat opens.

Sudden loss of oil pressure could be due to an oil pressure-relief valve sticking open, or, in aircraft equipped with oil-dilution systems, to a malfunction or inadvertent actuation of the dilution system, which reduces the viscosity of the oil by flooding it with fuel. Abnormally high oil pressure is unusual; it could be due to the use of oil with too high a viscosity number (most likely) or perhaps to a failure of the oil-pressure-relief valve to open. Sudden pressure fluctuations can mean that you are running out of oil − due to a rapid leak if a slow drop in pressure did not occur first and that the oil pump is beginning to pick up air. It could also mean that the pressure-relief valve is alternately sticking and releasing. A sudden and complete loss of oil pressure is usually an indication of a mechanical failure such as a broken oil line, failed bearing, or failed pump. These are all very rare − especially pump failure.

Oil temperature is closely related to oil pressure. Since the engine depends on oil for some of its cooling, oil temperature is a measure of a vital operation, especially in engines that are not equipped with oil radiators. However, a high oil temperature, as long as it remains within the green, is not a cause for alarm. In fact, oil temperatures that run consistently low, near the bottom of the green, are more deserving of attention. The reason is that elements other than solid dirt contaminate the oil. Filters remove dirt, but they do not remove liquid contaminants, which only boil off at high temperatures. Consistently low oil temperature may lead to incomplete boil-off of contaminants, which will then rust or corrode internal engine parts. Oil temperature tends to run consistently low in cold-weather operation, especially in engines not equipped with thermostatically controlled oil radiators. If consistent low-temperature operation is unavoidable, the oil should be changed frequently to get rid of the inevitable liquid contaminants.

Periodic oil-temperature fluctuations will be observed in engines equipped with oil-temperature thermostats, especially shortly after takeoff and during climb. These fluctuations indicate normal operation of the thermostat. Persistent fluctuation during flight or consistently low or high oil temperature in fair weather usually indicates a thermostat malfunction, although a sharply climbing indication in extremely cold weather may rather suggest blockage of the radiator by congealed oil.

High oil temperature may also be due to over or underfilling of the engine with oil, to excessively high power settings at low airspeeds, to fuel of too low an octane, or to laboring or lugging resulting from the use of high manifold pressures coupled with low rpm. In the summertime the air passages through the oil radiator can become plugged with insects, causing excessively high oil temperatures — a condition that a good preflight should preclude. Clogged and dirty oil-filter elements or screens, which retard the flow of oil through the system, also may produce high oil temperatures.

While the oil-temperature gauge does provide valuable information about the thermal balance of the engine, it is comparatively insensitive to rapid changes in temperature in some areas, such as the tops of the cylinders. In order to keep the pilot informed of temperatures in the cylinder heads, a temperature sensor is embedded in one of the rear or middle cylinders of the engine — the one that is assumed to be the hottest-running of all the cylinders. The cylinder head, however, is not the hottest-running part of the engine: the exhaust valves and stacks and the spark plugs run considerably hotter. Nevertheless, the temperature of the cylinder heads can give important information about more than the heads alone.

An abnormal cylinder-head temperature indication is usually on the high side. An uncommonly low reading might indicate insufficient winterization, open cowl flaps, when they should be closed, or simply insufficient power to keep temperatures up. High temperatures, however, are the important ones. An abnormally high cylinder-head temperature may indicate that the flow of cooling air over the engine is somehow being impeded. It may merely be a matter of too much power, too little airspeed, and too warm a day, as in a long, steep climb in hot weather. In cold weather however, ice can collect on the cooling inlets, constricting airflow. In spring and summer, it is not uncommon to find birds' nests inside the engine cowling on top of the cylinders. Anything — even a scrap of paper blown into the cooling baffles — that impedes airflow around the engine will cause a rise in cylinder-head temperature. At the other extreme incorrectly installed or missing baffles, which may permit too unrestricted a flow around the engine, can prevent efficient cooling and lead to a rise, not a drop, in head temperature. If the cooling airflow is working as intended, the cause of heating is internal. High readings at cruise power just after a fuel stop could mean that you have fuel of too low an octane for your engine. Excessive temperatures at cruise power immediately after a 100-hour inspection or engine check could indicate improper magneto timing. High manifold pressures combined with low rpm — or any other practice conducive to detonation — will cause high cylinder-head temperatures; operating with an excessively lean mixture is perhaps the most common example. Finally, anything that interferes with the free discharge of exhaust gas will also cause a rise in cylinder-head temperature.

The fuel-pressure gauge can also forewarn of trouble if the pilot is sufficiently familiar with his fuel system to interpret its indications. On engines equipped with carburetors the pressure gauge is used primarily to indicate that fuel pressure is within the desired operating range. A sudden loss of all fuel pressure followed by return to normal with activation of the auxiliary pump usually indicates a broken fuel line. Under these circumstances, the auxiliary pump should be shut off immediately and the fuel valve set at the "off" position: otherwise there is a possibility of fire. Fluctuating indications point to a depletion of fuel supply to the pump, which could be caused by a tank running dry or by leaks or obstructions in the line from the tank to the pump. Upward fluctuations may also indicate an obstruction in the line from the pump to the carburetor: in this case, a sudden high rise in fuel pressure is usually coupled with a loss of power. A sudden drop in fuel pressure coupled with a loss of power usually reveals a leak between the pump and the carburetor. If these symptoms are followed by a gradual drop in oil pressure, the probable cause is a ruptured diaphragm in the engine-driven fuel pump (which then lets raw fuel into the crankcase, diluting the oil). If a sudden drop in fuel pressure coupled with loss of power and engine roughness is alleviated by leaning the mixture, the problem is a float needle valve that has stuck open in the carburetor. If in cold weather you experience a gradual drop in fuel pressure and subsequent loss of power and the situation is not remedied by the auxiliary fuel pump, you can be reasonably sure that the cause is water contamination in the fuel tanks or lines, which is freezing somewhere between the tanks and the pumps. This possibility makes it imperative to drain sumps thoroughly when taking an airplane out of a hangar for a winter flight or whenever flight in below-freezing temperatures is planned.

The fuel-pressure gauge on a fuel-injection engine is quite another matter. This instrument is intended to monitor fuel flow with respect to power settings. In supercharged or turbocharged engines the subject becomes even more complicated. Troubleshooting with the fuel-pressure gauge on injection engines requires an intimate knowledge of the injection system and its operation, which is beyond the ken of most laymen and beyond the scope of this chapter. The other simple engine-monitoring instruments, however, may be very useful both for interpreting an incipient emergency and for preventing one as long as the pilot understands the function and significance of the gauges and the systems about which they speak well enough to make sense of their reports. For his own safety and for the good maintenance of his engine every airplane owner should familiarize himself sufficiently with his engine's entrails to make the readings on his instrument panel more than merely monotonous mumblings in an unknown tongue.

14.

SWINGING YOUR COMPASS

Does your plane have a built-in crosswind? It's probably due to compass error: most general aviation aircraft have compass deviations of 12 to 18 degrees on some headings. Pilots still get to their destination, largely because of radio navigation, but the process would be much easier with a well-calibrated compass.

The pilot can "swing" (i.e., calibrate) his own compass and get a really reliable calibration card whenever he wants to. Compensating it (that is, adjusting it) by turning the little N-S and E-W compensators in the compass housing is called *minor maintenance*; to be legal, a licensed A&P should supervise the work and make an entry in the logbook. Never adjust the compensators unless you are going to swing the compass immediately, since unexpected deviations may result that can cause all sorts of navigational trouble.

For compensating and/or swinging the compass, secure a Suunto KB-14 hand compass. (It costs about $17, and several suppliers are listed at the end of this chapter.) This is the only currently available hand-held compass with the necessary accuracy. It's faster than a tripod compass and allows you to ground-swing an airplane compass with the radios either on or off in 20 minutes flat. Before compensating or swinging the plane compass take a few practice shots with the Suunto hand compass to be sure that you can use it quickly and accurately. Since this is a very accurate compass, certain precautions are in order to maintain that accuracy. Take off your wristwatch and steel-rimmed sunglasses if a check shows them to be capable of changing a compass reading. If the air is cool and dry and you've been carrying the compass in the pocket of a wool shirt or otherwise causing friction against fabric, the plastic face of the will have picked up a little static electricity, which may attract the compass needle upward and cause it to stick. This condition is easily remedied by exhaling onto the plastic face of the compass.

Now take a practice bearing on some vertical line, such as the edge of a telephone pole or the corner of a building. When you want an accurate reading, it helps to borrow a trick from the expert rifleman: take a deep breath, then exhale half of it just before you steady down for the shot. With a good eye you will be able to estimate the reading to one tenth of a degree, and you will be within two or three-tenths of a degree of the correct magnetic bearing most of the time. After a few practice shots you're ready to ground-swing your airplane's compass. If you're fairly sure that its compass is not badly out of adjustment, you can skip the adjustment procedure and merely calibrate the compass without adjustment.

If you decide to adjust the compass, select a flat area free of magnetic devices. A level field or parking apron is fine if there are no large bodies of iron or DC electric cables underground or overhead. While the job can be done alone, you should have a friend to man either the plane or the external compass. Crank up the engine, turn on the radios, taxi onto the level area on a bearing of 0 degrees, run up the engine, and read the plane compass. (This instruction assumes that you have a tricycle-gear airplane, for the compass must be in a level-flight attitude; if you have a taildragger, you must block up the tailwheel with a stool or sawhorse before each reading.) While this is being done, the external man kneels about 50 feet behind the plane, adjusts his position until he can simultaneously sight on the center of the tall skid and the center of the nosewheel strut with the Suunto compass, and takes a reading, which is the actual magnetic heading of the plane. Mr. Outside runs up and reports the mag heading to Mr. Inside, who then adjusts the N-S compensator so that the plane compass will read that magnetic heading when he runs up the engine and rechecks it. Mr. Inside then turns the plane to a heading of 90 degrees, rolls it ahead a bit to straighten the nosewheel, and again runs it up and reads the compass, while Mr. Outside sights in the magnetic heading of the plane as before and reports it. Mr. Inside now adjusts the E-W compensator until the plane compass reads the reported mag heading on subsequent run-up. He then turns the plane to a heading of 180 degrees and again runs up and reads the compass, while Mr. Outside sights the mag heading and reports it. Now the routine changes. Mr. Inside adjusts the N-S compensator only enough to remove half of the deviation. That is, if the plane compass heading had read 179 degrees on run-up and Mr. Outside said the Suunto gave a mag heading of 183 degrees, then Mr. Inside would adjust the compensator so that the plane compass would read 181 degrees on subsequent run-up. Then he turns the plane to 270 degrees and repeats the run-up and reading procedure, and again, on this westerly heading, he adjusts out only half of the deviation. The compass is now compensated for permanent magnetism and is ready to be swung.

While it seems easiest to describe the compensation procedure in the north-east-south-and-west sequence, you can start on any of the four cardinal headings and proceed through the other three in sequence as long as you remember to adjust all the deviation out of the first two adjacent cardinal headings and only half of the remaining deviation out of the last two. If the first one or two cardinal headings that you check are off 2 degrees or less, skip them and start your compensation with the first cardinal heading that shows a larger deviation. Once you start compensating, however, continue on around and compensate (or at least check) the remaining three cardinal headings. You may decide that it's worth the trouble to check all four cardinal headings before compensation and to start the procedure on the heading that has the largest deviation.

Now for the swinging process: the ground swing goes faster if each person has a data sheet and records his data independently. Mr. Inside should have five columns on his data sheet, captioned "magnetic heading, deviation, compass heading, deviation, compass heading." Columns two and three should have a superior caption of "radios on"; columns four and five, "radios off." Mr. Inside warms up the radios while taxiing to any heading that is a multiple of 30 degrees. He then runs up the engine, reads the compass when it steadies down, and records the compass heading in the "radios on" column. He then shuts off all the radios, runs up the engine again, and records the compass heading in the "radios off" column. While he is doing this, Mr. Outside is sighting the longitudinal axis of the airplane with his Suunto and recording it in his own magnetic-heading column. Mr. Inside turns his radios on to warm them up while he is taxiing to a heading of 30 degrees, where he again reads and records his compass heading for both "radios on" and "radios off." This procedure is repeated for each 30-degree increment of heading around the full circle, and Mr. Inside and Mr. Outside both record their headings at each run-up. (If the radios warm up slowly but the plane is small and turns quickly, it may prove quicker to swing the compass clear around the circle with the radios on all the time and then repeat the swing with the radios off all the time. Either way you do it, however, it is important to get a radios-off swing. You may need accurate dead reckoning the most if your radios are out.)

With the above figures, transfer Mr. Outside's magnetic-heading figures to Mr. Inside's data sheet and calculate the compass deviation by subtracting the corresponding magnetic heading from each compass heading. If the magnetic heading is larger than the compass heading, be sure to place a minus sign in front of the difference to show that the deviation should be subtracted from the mag heading to get the compass heading. You now have a deviation value for "radios on" and one for "radios off" for each magnetic heading. Plot these values on graph paper with the magnetic heading reading horizontally from 0 to 360 degrees through the middle of the page and with the deviation plotted above (+) and below (-) this line to some convenient scale. Make a separate plot for the "radios on" and "radios off" values by plotting the "radios on" values in black and the "radios off" values in red.

You will use these curves to determine the values for a new compass-deviation card but also make up this graph with large, easily-seen printing and fasten it to the back of your sun visor or keep it in your flight bag. Since fatigue and stress can make your mind play tricks, you might print in bold letters on the graph "Add deviation to magnetic heading to get compass heading." With the above calibration you probably have a better compass card in your plane than you've ever had before. Go out and do a simple dead-reckoning or radio-navigation problem. You'll be amazed at how that built-in crosswind has vanished. If you have any drift, it's because a real wind is pushing you around.

Publisher's Note: The Suunto compass is apparently no longer available. However it seems clear that any quality compass capable of one degree or better resolution will certainly be adequate.

15.

THE INSTRUMENT THAT (USUALLY) ISN'T THERE

The airspeed indicator tells airspeed and nothing else, but when we use it in slow-speed flight as a device to measure lift coefficient — a purpose to which it is adequately, but not excellently, suited. Lift coefficient is a figure used to describe the behavior of wings: it relates lift to speed in such a way that for the lift to remain constant, the lift coefficient must change in inverse proportion to the square of the airspeed. In other words, if you double your speed, your lift coefficient goes down by a factor of four. The practical range of lift coefficients for normal wings in normal flight is from 0.1 to around 1.6 with no flap or perhaps 2.5 with a slotted flap fully deflected.

Lift coefficient is of no practical interest to the pilot except in that it has an upper limit that limits the lowest speed at which an airplane can be flown. It happens also that besides the stalling speed, several other important speeds — best angle and rate of climb, best glide, best endurance, approach and slow-approach speeds and best range speed — can be mapped on a scale of lift coefficients. By relating them to lift coefficient, they are made independent of airplane weight and therefore more relevant to actual flying conditions. For example, the book best-rate-of-climb speed applies only at gross weight: at light weight the best rate of climb is at a lower speed. The handbook usually does not give a range of reference speeds for different weights, and the pilot is left to apply as well as he can the rule of thumb that all the reference speeds go down with a decrease in weight.

If airspeed is not directly related to lift coefficient, we must find something else that is in order to have a true measure of the conditions under which the wing is operating. At first glance, angle of attack would seem to be it. Angle of attack is for all practical purposes the angle between the fuselage reference line and the direction of flight. The proper reference line is strictly the chord line of the wing, but the fuselage reference line is easier to visualize and keeps us out of a confusing thicket with deflected flaps.) In order to measure it, we have only to put a vane out somewhere in undisturbed air, let the vane pivot and drive a potentiometer, and read the output on the panel as a measure of degrees of vane deflection. This will indeed work well, and it will produce the gratifying result that the airplane will always stall at the same indicated angle of attack, whether it is moving fast or slowly, is light or heavy, is pulling G or no G, banked or level. There's only one hitch: As soon as you deflect the flap, you equip the airplane with a new wing, which in effect is attached to the fuselage at a different angle of incidence than the old wing. We now find that the stalling angle of attack has changed — although as long as you stay in this configuration, the airplane will always stall at the new angle of attack.

Since the airplane may have any number of possible flap settings as well as a couple of gear positions that also may affect stalling attitude, interpreting the vane becomes a rather complicated matter. Perhaps an electronic network could be installed between the vane and the panel meter to shift the meter reading around in such a way as to compensate for different flap settings. This is what is done in the systems found on some Learjets, in which the vane is mounted on one side of the nose close to the fuselage skin. (Since the meter can be calibrated to make up for position error in the sending unit, it is not actually necessary to locate the vane in undisturbed air: any position in which flow direction varies consistently with angle of attack will do.)

Another type of sensor completely bypasses the vane and the angle of attack of the airplane and measures pressure patterns at the leading edge of the wing. This type is identical in outward appearance to the small, square stall-warning sensor on the leading edges of most general-aviation airplanes, but, rather than containing an on/off switch, it contains a type of potentiometer and a spring-balance arrangement that lets the sensor blade (which, confusingly, is also called a vane) move forward or aft in reaction to pressures around the leading edge. Again, the potentiometer operates a panel meter, and the meter is calibrated in terms of certain standard reference speeds. The leading-edge vane acts just like an idealized lift-coefficient sensor; in most cases, it is self-compensating for flap movement and is unaffected by gear and power settings.

The only device of this type that is now on the marker is made by the Safe Flight Instrument Corporation, the same company that makes all the leading-edge stall-warning sensors found on lightplanes. Lift sensors have been in use on sophisticated airplanes for years, but though Safe Flight offers models for light aircraft at under $500, it is rare to see one.

The panel meter, which is mounted on top of the glareshield, has an edgewise display about 1/2" high and 2 1/2" wide. The meter was designed to be mounted flat, with the slow end of the band to the left, represented by a red zone, and with a yellow arrow giving the reading. The editors probably did not give ourselves enough time to get used to this display, but at first we found it a little difficult to interpret rapidly: does "left" mean "up" or "down," for example? Accordingly, we had the meter reinstalled vertically, with the stall zone at the top. The people at Safe Flight thought this arrangement backward; they do supply some instruments for vertical mounting, but they put the stall at the bottom. The difference lies in the assumption made about the pilot's interpretation of the display. The editors saw it as an artificial horizon, with a bug — the moving arrow — that represented the airplane: "up" meant "nose up," as it does on the attitude gyro. Safe Flight's customers, however, regarded the arrow as a flight-director bug; a down movement of the bug meant "fly down," which would have the effect of decreasing the lift coefficient and bringing the bug back upward.

We think that for light-aircraft applications the artificial-horizon analogy is better. Most of us are not in the habit of following a flight-director bug, and, though we do have the habit of following a glideslope needle in which a low needle means "fly down," there is no analogy between the lift-sensor arrow and the glideslope needle: the needle represents a reference point in space toward which you are flying, while the arrow represents a condition of the aircraft that you are directly affecting by your control inputs. The display consists of a small vertical rectangle with the red stall zone at the top, marked by the word "slow" (which happens to be written sideways, since the Sierra's meter was meant to lie flat). At the bottom is the word "fast," and across the middle is a cross-hatched line (called the "barber pole" for no obvious reason). At about the 1/4- and 3/4-scale points are two diamonds: the upper one is called the slow diamond, and the lower one the fast diamond.

Getting the sensor adjusted properly, even with the help of Safe Flight's opaque instruction manual, is quite a trick: ours is still not right after several hours of trying, but it is close enough to be useful. They had one on a previous airplane — a Lark Commander — that they never managed to adjust at all and finally returned to the factory. It was an older version of the instrument, however, and the present one represents a quantum improvement in serviceability. Once it is working, it virtually replaces the airspeed indicator except for measuring cruising speed. Its stall-warning function is tested before takeoff with a button on the meter. (The audible stall-warning device is now incorporated in the lift-coefficient meter.) After takeoff, if you want best-angle-of-climb speed, you put the arrow on the slow diamond; for best rate of climb, it goes on the barber pole. The fast diamond gives a good cruise-climb speed. The sensor self-compensates for your flap setting, if any. During cruise you forget it, but at the other end of the flight you slow down to the fast-diamond speed once you are on the ILS or as you are about to enter the traffic pattern: this is your maximum flap-operating speed. When you lower the landing gear or are starting your pattern proper, you slow down to the barber-pole speed. In the Sierra this is around 81 knots — the approach speed. Then a little miracle happens: as you add flap, the IAS corresponding to the barber-pole position drops until, with full flap, it is the recommended full-flaps approach speed — 74 knots in the Sierra. No trim change is needed throughout the approach. What is happening is that lift coefficient is changing, since you are slowing down as you add flap, but the percent of maximum lift coefficient is remaining about the same. For the final phase of the landing, the lift indicator is ignored, although, if you feel like it, you can follow it throughout the flare, because it is so narrow that, when you focus your eyes on the runway, the two images of the meter do not overlap. They are both transparent but readable (though a little blurred).

A short-field approach is flown in the same way but with the arrow on the slow diamond throughout the approach. This system, of course, compensates for weight. Whether you are alone in the airplane or you've brought your family and friends, you fly in exactly the same manner. You can see the weights reflected in the IAS: at the barber-pole best rate of climb the Sierra indicates around 77 knots if it is very light, and 81 at gross.

That a few little marks on a dial manage to stand for every characteristic speed of the airplane is mainly a matter of luck: the speeds do in reality cluster around a few points, and some of them, such as approach speed, have no strict definition and can be defined in terms of a more specific speed, such as best-rate-of-climb speed. The meter defines approach speed, for instance, as the indicated speed with gear and flaps down when the meter indication is the same as that for the rate of climb with the airplane cleaned up. On the Sierra, all these speeds come out remarkably well; even the maximum flap-operating speed, which is an

FAA specification rather than an aerodynamic one, coincides exactly with the fast diamond. If one or another of the speeds did not line up perfectly, however, you would have only to make a mental note of the discrepancy and fly accordingly: for instance you might prefer to make your slow approach with the arrow one diamond width to the fast side of the slow diamond.

If this device were nothing more than a glorified airspeed indicator, it would be of little value. What makes it tremendously helpful is its independence of every sort of disturbance and the high natural stability of the information that it delivers. To a degree exceeding anything that the author would have imagined, with the Safe Flight, you find yourself cruising down the glidepath like a car down a turnpike. Practically no flying is necessary. You simply ignore the airspeed indicator except to notice its surprises − for instance, that you really have to drop the nose and pick up speed on your turns to keep a constant margin above stall. Because of the greatly increased precision of the approach the final flare and landing are apparently better as well: smoother and more accurate even on the Sierra's rocky rubber-doughnut landing gear. Climb performance is also enhanced.

The essential trick is that the lift indicator gives you real-time angle-of-attack information. The airspeed indicator is always lagging, because the angle of attack must first decrease (for instance), and the plane then accelerate to a new trim speed before you become aware of a change. By the time you correct, it is too late. With the lift indicator you pick your angle of attack and let the speed catch up on its own time. You always fly at the most efficient angle of attack for the phase of flight that you're in. Any pilot, even one with a high level of proficiency, would find a lift-coefficient indicator an invaluable help in getting the most out of his airplane. Students might also find the entire approach-and-landing process substantially easier if they had a real-time display to work from rather than the airspeed indicator. For everything except cruising it is *the* pitch-control reference. It also has an advantage over the conventional audible stall warning in that it gives a progressive display of approach and entry into the stall region as well as an audible warning of the coming stall break. The supplementary detail in stall warning is a help, but it should not be used by manufacturers, as the simpler tab-type stall warner has been, as an excuse to overlook a good aerodynamic stall-warning system in their airplanes.

16.

HOW RADIOS WORK

Publisher's Note: This chapter was written before Loran and GPS had become functional and valuable air-navigation tools. Thus, the information presented is accurate, but incomplete to that extent.

If you do not like knowledge for its own sake, stop here, because this is the only sentence in this chapter that has any application. We are going to talk about how radios work, not how to work them nor how to repair them nor how to afford them. We only want to satisfy your curiosity.

Radio waves are part of the electromagnetic spectrum, which includes radioactive gamma rays, ultraviolet, visible light, infrared, and longer and longer wavelengths. Electromagnetic waves differ from ocean waves and sound waves in that they do not need a medium through which to propagate; rather than vibrations of matter, therefore, they are oscillations of magnetic and/or electric forces. The whole subject borders upon the metaphysical, in fact, so let us leave it.

Like sound and light, radio waves may be used for communications. They have the advantage of long range: while sound dies quickly and is easily blocked — and light, too, is blocked even by fog or haze — radio waves of various types can penetrate the atmosphere and many solids and, by using atmospheric reflection, can even be directed around the curvature of the earth's surface. Visible light is prone to interference and very long-wave radio is too static; in between the two in the UHF and VHF (ultra high-frequency and very high-frequency) ranges radio waves can be beamed like light, can penetrate fog, and are easily produced, managed, received, and decoded.

Electrons, especially in metals, tend to oscillate in time with radio signals. An antenna is a container of electrons that vibrate in tune with passing waves, generating electric currents — by definition, movements of electrons — that can be detected by the radio receiver. The receiver can be adjusted to receive oscillations of only one frequency at a time; each sending station has a "carrier" frequency — a basic fixed rate of electron oscillation — which can be located and isolated by a receiving set. Information is then superimposed on the carrier wave in the form of "modulation" — small changes in the power (amplitude modulation, or AM) or frequency (frequency modulation, FM) of the signal. The receiving set separates the modulations from the carrier signal and then does useful work with them — amplifies them into an audible signal, for instance, for voice communications, or uses them to drive cockpit indicators and black boxes for navigation. If the tower tells you that you have "carrier but no modulation," he means that he is getting a hum or hiss that indicates that you have keyed your mike and are beaming out a signal but that the signal lacks the modulation necessary to communicate information.

It is easy to tell where a visible light is coming from. To tell where a radio beam is coming from is more difficult, but it can be done in several ways, one of which is used in the venerable piece of navigation equipment called the automatic direction-finder.

The trick involves using a loop antenna that is able to rotate. The axis of rotation is at right angles to the plane of the arriving radio signal. For any position of the antenna (except if the plane of the loop directly faces the radio source) one side of the antenna will be slightly closer to the sending station than the other. Since radio waves are waving up and down as they go along, the wave has a slightly different height − i.e., power − when it hits the leading side of the loop than when it arrives at the trailing side. If the wave striking the leading side is at a higher point on its oscillation than when it reaches the other side, there is a tendency for electrons in the loop to flow from the front edge toward the back edge. By rotating the antenna you can change the net direction and magnitude of flow until, when the plane of the antenna faces the sending station, there is no net flow. A pointer on the instrument panel is slaved to the loop, and both come to rest when a "null" signal is felt by the loop. The ambiguity about the direction of the signal (it could be coming from in front of or from behind the loop) is resolved by a supplementary linear antenna. In most modern ADFs there are two or three fixed loops instead of a rotating loop and, by comparing the signals in them, the receiver computes the direction to the station electronically.

OMNI

The radio waves used by ADF are isotropic; that is, the same signal is broadcast in all directions. Other systems, such as the old long-wave radio ranges and the modern omni network, use signals that are direction-dependent or, in the case of loran, position-dependent.

Omni stations broadcast their signals in all directions, but the signals vary around the compass in such a way that each direction has its own signal, which cannot be confused with that of any other direction. If a receiving station can pick up and decode the signal from an omnirange station, it can tell its bearing (or radial) from the station. As a convenience the omni head can add 180 degrees to the "from" determination and instruct the pilot which way to fly "to" the station: the to/from flag tells the pilot in which mode the head is operating.

The VOR station works its magic by modulating its carrier wave with a 30Hz (cycles-per-second) signal the phase of which depends on direction. As an analogy, suppose that the VOR station is aiming two spotlight beams at the aircraft, one orange and the other blue. Both beams are amplitude-modulated − that is, they regularly grow bright and dim at a fixed frequency − at, say, one bright/dim cycle every minute. The lights are controlled so that, when they point north, they are in phase, and, if the pilot sees the light growing bright and dim in step, he knows that he is due north of the station. Suppose that the lights are controlled so that the blue light's timing stays constant around the compass, but the phase of the orange light is directionally dependent. As the lights are turned to the right, away from north, the orange light begins to lead the blue one. At a position 30 degrees right of north — or, in omni language, on the 30 degree radial — the pilot will see the lights getting bright and dim once per minute as before, but now the orange light begins to brighten 5 seconds before the blue and correspondingly to dim 5 seconds sooner. At 90 degrees, the orange would lead by 15 seconds, and at due south by 30 seconds — that is, when the blue light was dimmest, the orange one would be brightest. In such a system the pilot can determine his direction from the station if he can measure the phase lead of the orange light. Omni works in the same way. A 30-Hz fixed-phase modulation is broadcast in all directions, and another 30-Hz signal is transmitted with a directionally dependent phase lead. The "orange" is distinguished from the "blue" by a tricky combination of amplitude and frequency modulation. The omni receiver merely determines how far the variable-phase signal has to be shifted back in time to fall into step with the reference signal and turns this information into a radial designation for the pilot.

PULSE EQUIPMENT

While omni gives accurate, specific directional information, it cannot make explicit distance measurements, though the pilot may find his distance from the station by taking an intersection of radials of two omnis or by doing a timed radial-crossing maneuver with a single omni. There do exist electronic instruments for measuring distance with radio waves, however: radar is one and DME is another, and in a real sense they are the same thing.

A radar transmitter on the ground sends out a short burst of carrierlike waves called a pulse. This transmission, the initiation and duration of which are exquisitely timed, lasts about a millionth of a second at a typical frequency of several billion cycles per second. The pulse of the radar is analogous to a blink of a flashlight. The narrow beam of energy is fired off in a direction that is accurately known from the orientation of the antenna at the moment of sending.

When the pulse strikes a solid object, such as an airplane (or even nonsolid objects such as precipitation), part of it is reflected, just as a flashlight blink may momentarily illuminate an object. Part of the reflected signal, still identifiable by its characteristic frequency and duration, returns to the antenna, which is now idle between pulses, and is passed on to the radar receiver. The receiver notes the direction from which the reflection came, again by the antenna orientation, and it measures the time delay between the sending and the return of the pulse. It displays this information, along with similar information for other reflections, in maplike form on a cathode-ray tube similar to a TV tube. Since radio waves travel with a fixed, known velocity of about 186,000 miles per second, the measured round-trip time for a pulse is easily converted into a distance. Radio waves require 12.36 microseconds for a 2-mile round trip, so a delay of 123.6 microseconds would indicate a target 10 miles away.

The cathode-ray-tube display is arranged with the antenna at the center of the screen and with the targets surrounding it in positions determined by their measured distance and direction. Concentric circles showing mileages from the station and outline maps of fixed landmarks may be permanently imprinted on the scope to aid the controller in visualizing his targets.

Most ATC radar is of the surveillance type, which gives the direction and distance of the target but not its altitude; a more sophisticated type of radar, infrequently encountered, measures not only the compass direction of the target but also its elevation above the horizon and combines elevation and distance to compute altitude. The radar transmitter emits about 1,000 pulses every second (thus it is idle 99.9% of the time), but its antenna typically rotates at about 10 rpm, and a given aircraft will encounter a burst of pulses every six seconds or so when the antenna is pointed in its direction − hence the intermittent blinking of the reply light on a transponder.

TRANSPONDERS

During World War II, while radar was in its infancy but was already changing the course of the war, it became apparent that it would be useful to be able to distinguish between friendly and enemy targets. IFF (Identification Friend or Foe) was invented to fill this need, but the basic device, now known as a transponder, has become essential to civil aviation as well as to military aircraft. The transponder is a small airborne transmitter that waits until a radar pulse strikes its antenna and then instantly broadcasts, at a different frequency, an isotropic radar reply pulse of its own − a strong, synthetic "echo." Since ordinary "skin return" − the reflection of the ground-radar pulse from the airplane structure − is sometimes quite weak, especially at great distances and in a radio-noisy environment, the transponder enables the radar operator to track targets that might otherwise return too weak an echo to see.

A transponder is simple in concept but in practice is a complex, sophisticated device. It is triggered into either of two modes of reply by the nature of the ground-radar pulse. If the interrogating radar sends two closely spaced pulses 8 microseconds apart, the transponder replies in the normal mode-A identity code. If the interrogating pulses are 21 microseconds apart, the transponder comes back with a mode-C reply, which means that the altitude of the airplane, taken from a specially equipped altimeter, is encoded in the reply to appear in numerical form as part of the target showing on the controller's scope. In its reply code, the transponder sends a sequence of 0.45-microsecond pulses spaced 1.5 microseconds apart. The timing of the sequence and the number of pulses in it codes the desired information, including the numerical code (0700, 1200, 7600, etc.) in which the ground controller instructs the pilot to reply, or "squawk." By limiting his scope's display to the numerical codes that he is interested in, the controller can block out returns from aircraft that are being worked by other controllers. The transponder replies appear on the controller's scope as pairs of parallel dashes or hash marks; when a transponder squawks "ident," a third hash mark appears briefly between the other two, brightening the image and making it stand out clearly for identification.

In ARTS (automated radar terminal service) radar replies are channeled into a computer that decodes the pulses, converts them into letter-number displays, and places an identifying sign, including the number and altitude of the aircraft, next to each target on the radarscope.

DME

Another application of the radar-transponder system but in reverse is distance-measuring equipment. In this system the aircraft carries the radar sender and the ground station in the transponder. The airborne DME sends out carefully spaced pairs of pulses at random intervals, then listens for the characteristically spaced replies, or echoes, from the ground station. Since the ground station may be replying to interrogations from several DMEs at the same time, each DME must somehow sift its own echoes out from among the others. It does this by remembering the time of all the replies that it receives and sorting through them for the ones that occur at about the same time intervals after each pair of its own interrogating pulses. When it identifies a set of pulses as its own echoes, it follows only its own replies. Like ordinary radar, the DME simply measures the transit time in which its pulses reach the station and return and converts the information into a distance. Rather than display the information on a scope, however, it displays it digitally or on a meter face. Many DME sets also take the additional step of integrating the rate of change of distance and presenting the result as a groundspeed indication. This indication represents the speed of the aircraft only with respect to the ground station, thus it gives a true groundspeed only when the aircraft is tracking a radial to or from the station.

RNAV

Just as the DME presents a groundspeed measurement arrived at mathematically rather than from observation, an area-nav (RNAV) set presents the aircraft's position with respect to imaginary omnis and "waypoints" set up by the pilot. It does its work by trigonometry, combining known angles and distances to find the locations of unknown points. Just as you could, if armed with a chart and a couple of omni bearings, find your position and then see which radial of a third omni you are on, the area-nav computer determines your position with respect to a third omni that doesn't really exist. In effect, all it is doing is charting your position in terms of existing omnis and then presenting it to you as if you were tuned to an imaginary omni that is probably placed directly upon your intended course. It cannot synthesize an omni without having two real omnis or one plus DME to work with, however: if real omnis are out of range, an imaginary one is also out of range, even if it is right beneath you − which brings the subject back to metaphysics and a good place to stop.

17.

INSIDE AVIONICS

Avionics are frustrating, fickle, and difficult to understand but essential to much of what we pilots want to accomplish. So we learn to live with their shortcomings and appreciate their assets. Radios are often the tail that wags the dog. Navcoms become our masters; they govern, they dictate, they determine which aircraft we fly and whether or not it is operational. We are pilots, not electrical engineers, so our dependence on invisible electrons playing tag in some incomprehensible black box compounds the frustration.

Life was easier in earlier days: you bought the best, most powerful, most versatile radios around, and if you were budget-limited, there wasn't much selection. In the mid 1950s Ed King was still working for Collins; Bill Rice of Genave was earning his spurs at Hazeltine Labs; and the transistor had yet to be perfected for use in avionics. ARC was a privately owned firm that specialized in high-quality, moderately expensive corporate equipment; Bendix was synonymous with airline avionics; and Edo's contribution to general aviation consisted mainly of seaplane floats. It was pretty much Narco or nothing.

Great strides have been made since then, however — and living with avionics has become even more frustrating. At least in the 1950s, the pilot just took what he could get. Now the greater selection of hardware plus the increased reliance on navcoms, transponders and DME for all aspects of flight — VFR as well as IFR — has complicated the problem, and the radio requirements for flying locally VFR with limited traffic are vastly different from the avionics needed for extensive IFR cross-country and high-density terminal-area operations. Pragmatic or not, a pilot is forced to know more about avionics in order to cope in a world dominated by specifications that are foreign to his vocabulary, by regulations, by TSOs and MOSes, by reliability that at times appears quite elusive, and by that queasy feeling that comes from being dominated by a force that you do not fully understand.

Examining specifications and performance figures seems to be a reasonable way to learn about a piece of avionics — provided that you can interpret the meaning of the various specs. Even a simple item such as transmitter power output needs some explanation, since it is an oversimplification to say that the larger the rated power, the better the radio. Moreover, the amount of power may be incorrect or at least inappropriate for certain circumstances: the Radio Technical Commission for Aeronautics (RTCA), a leading force in establishing guidelines for government avionics regulations, states that a VHF transmitter shall not exceed 25 watts of effective radiated power except under special circumstances requiring extended-range operations. In essence, the RTCA is suggesting that transmitter wattage has a practical upper limit beyond which the range and interference-penetrating advantages of increased power lose relevance even for airliners and corporate jets, which must communicate over much longer distances than a Cherokee or a Bonanza. Their studies indicate that 1 watt of radio-frequency carrier power − the energy unit usually given in the transmitter specifications − at the transmitter output terminals is sufficient to reliably produce adequate transmission at line-of-sight distances of 50 nautical miles and that 4 watts are sufficient for distances of 100 nm, and 16 watts are good for 200 nm, considering normal installation and antenna losses.

The RTCA studies emphasize that the desired level of transmitter power output depends upon the nature of the operation. If your flights are limited to the terminal area or to routes that require relatively short communication distances, your power-output needs can be quite modest. If a lot of other pilots are using the same frequency (such as a unicom) or if there is another form of interference on the channel, extra power is nice to have: the more power output, the better the chances of being heard over the competition. Notice that the signal strength increases as the square root of the transmitter power, so you must up the power output four times to get double the signal strength at the receiving antenna. The important element for reception is the electrical power being radiated from the antenna, so an improperly installed high-wattage transmitter will be no better than a less powerful unit with an efficient antenna installation.

The term *peak envelope power* (PEP) is occasionally listed with the power-output specs. This refers to the maximum, or, peak power of the modulated carrier wave. This wave does what its name implies: it carries the modulation, or voice characteristics, of the transmission. If the peak modulation of a 2-watt carrier wave were 100%, the PEP would be 8 watts; if the modulation were 85%, the PEP would be 6.74 watts. Several features of the modulation process are noted in transmitter specifications. One is the percent of modulation. The higher it is, the higher the peak envelope power will be. According to the RTCA's minimum operating specifications, modulation should be at least 85%; it should not exceed 100%, however, or distorted, fuzzy transmission will result. Another feature to look for is the way in which the transmitter prevents over-modulation. Shouting into the microphone, for example, would normally cause over-modulation. Many radios employ an auto-limiting, or clipping, technique that clips off strong voice signals that would over-modulate the carrier wave. More sophisticated transmitter designs, such as the Collins Low Profile VHF-20 and the Bendix TR-241 A, employ a gain control, which compresses the peaks of the voice signal that otherwise would create over-modulation. Both techniques are acceptable, but the compression method tends to be a more desirable, more pleasant way of preventing distortion.

The importance of frequency stability for a transmitter really depends upon pending government regulations. Maximum variation of +/-.005% in the frequency output from the specified channel or center frequency is now required but there is a notice of proposed rulemaking that may require +/-.003% stability. The higher degree of stabilization will be required for 25 kHz spacing between channels. Audio-frequency distortion is a transmitter characteristic of less significance, and it may or may not be included in a typical data sheet. It is the percentage of distortion and noise in the carrier output from the transmitter and in fact is a measure of the intelligibility of the transmitter. The RTCA recommends that it be kept to 25% or less; 10% to 15% is typical of transceivers such as King's Gold Crown KTR 900A and Collins' Low Profile VHF-20.

Sensitivity is a measure of a receiver's potential for producing clear sound and of the absence of background noise or hissing when signals are weak. It is expressed on most avionics spec sheets as the minimum input signal going into the receiver that will produce an output receiver signal − the sound you hear − that is twice the value of unwanted background noise. The lower the voltage given in the specification, the more sensitive the receiver. The RTCA's MOCs specify that a receiver should exhibit the base-line two-to-one ratio of output sound to noise with 10 uv or less of incoming signal. Typical values for most general-aviation avionics are 1.5 to 2.0 uv . These figures assume that the losses in receiving signal from the antenna to the radio are normal, since a poor installation can cause large losses that will significantly reduce the overall sensitivity of the receiver/antenna system.

Sensitivity is desirable, provided that the receiver has sufficient selectivity − the ability to receive a desired frequency without also getting interference from the immediately adjacent frequencies. Selectivity is presented as two sets of numbers. The first set is expressed as six dB (decibels) and some number of cycles per second, such as +/- 15, expressed as kHz, or thousands of cycles per second. The second set is 50, 60, or 70 dB and some other number of kHz, such as +/-40. Decibels are a means of expressing the ratio of two quantities such as two transmitting powers or two receiving-signal voltages. The higher the decibels, the larger the ratio and vice versa, but the scale is not linear: 60 dB is more than 10 times larger than 6 dB, and a voltage ratio of 60 dB is different from (smaller than) a power ratio of 60 dB.

The first set of selectivity specifications means that the ratio of the signal received from the desired frequency (122.8 mHz, say) compared with the signal received from a frequency 15 kHz above or below (e.g., 122.815 mHz or 122.785 mHz) will be two to one. Expressed another way, signals 15 kHz higher or lower than the selected frequency will be one-half as strong. The second set presents similar information but for a different ratio. Selectivity expressed as 60 dB +/-40 kHz, for example, means that a signal produced by a frequency 40 kHz above or below the desired frequency is 1,000 times weaker than the signal produced from receiving the selected frequency. Selectivity, therefore, is a measure of the width of the receiving band: the farther from the desired frequency, the less sensitive the receiver is to the adjacent frequencies Normally, you want a sharp bandwidth so that you receive only the selected frequency, thus it is desirable from a specifications standpoint to have small frequency values associated with the 6 dB and 60 dB ratios. If the specifications are for 50 or 70 dB rather than 60 dB, some allowance can be made for the fact that 50 dB is a smaller ratio than 60, and 70 dB is a larger ratio.

There are several other specifications that are somewhat similar in purpose to selectivity, although they refer to the rejection of signals and interference caused by frequencies far removed from the selected frequency. For example, spurious response is the receiver's response to or reception of undesired signals. Image rejection refers to the receiver's ability to reject − or not respond to − a special frequency, called the image frequency, that is generated internally within all heterodyne receiver designs. These specifications are given as so many decibels "down," which means that the receiver's response to these unwanted signals will be less than its response to or reception of the desired frequency by whatever ratio applies to the decibel rating given in the specifications. The larger the decibel rating, the larger the ratio, and ideally all responses except those generated by the desired frequency should be as many dB down as possible. The MOCs give desired minimum dB ratings for specific frequency ranges outside the normal VHF operating band.

The MOCs also state that the combined noise and distortion in the system audio output should not exceed 25% of the rated power of the output over the audio-frequency range of 350 to 2,500 Hz. This characteristic is called audio-frequency distortion. The range of sound that we are accustomed to hearing is between 30 and 10,000 Hz, and sounds at frequencies of up to 16,000 to 20,000 Hz are audible to some people. Aircraft radios, however, are designed to produce distortion-free sounds only within the relatively narrow range of frequencies produced by normal speech. The frequency range for a person's voice actually extends from about 100 to 10,000 Hz, but very little intelligibility is lost if good frequency-distortion characteristics are limited to a narrower range. For example, the distortion-free range of a telephone receiver corresponds closely with the 350-to-2,500 frequency band used in aircraft transceivers. A measure of how well the receiver produces sound is the audio-frequency response of the unit, which is expressed as a variation between the strongest and weakest audio output over the selected audio-frequency range. The maximum variation should be 10 dB between 350 and 2,500 Hz according to the MOCS, but the ideal receiver has what is called a flat audio response: it does not emphasize or slight any sound in the audio range.

Good specifications are not hard to find but beware: the more you want, the more you pay. For example, the cost of a transceiver is closely related to the transmitter power output. Each of the three different King transceivers costs about $155 per watt − the Silver Crown KY 195 ($1,090/7 watts), the Gold Crown KTR 900A ($3,050/20 watts), and the airline − Arinc-type KTR 9000 ($3,950/25 watts). The Collins Low Profile VHF-20 will run you about the same. Other factors in addition to transmitter power obviously enter into the cost of a particular radio, but the fact remains that more performance requires more sophisticated circuit designs and more powerful components. The market forces that require a Silver Crown KY 195 to have 7 watts and a Gold Crown KTR 900A to have 20 watts are the same forces that require the Gold Crown unit to have more features than its Silver Crown counterpart. If performance is achieved at the expense of taxing the design and its components to their fullest extent, impressive specs can cause persistent headaches. Edo-Aire, for example, chose to derate their new RT-553 navcom to 5 watts, even though the box was capable of putting out twice that power.

Receiver specifications are also related to complexity of design, components, and thus cost. Economical units such as the Geneva Alpha 600 take the signal from the receiving antenna and pass it through a filter and mixer so that all the amplification of the radio-frequency (RF) signal is achieved in the radio's intermediate-frequency stage. They do not employ an RF amplifier and tracked filter, which are found in receivers such as the Bendix 241 A and the Collins Low Profile VHF-20. In theory the use of an RF amplifier should provide better reception of low-level signals, but the sensitivity specs for all these transceivers are nearly identical. With respect to selectivity, the more impressive the specifications, the more the unit will cost, and the relationship of price to performance can soar as the selectivity or bandwidth values narrow. Omni tolerance is another spec for which the cost of achieving accuracies better than 1 to 1.5 degrees is considerably greater than the cost of obtaining less impressive yet legally acceptable levels. The design dollars that influence VOR accuracy, however, dictate only about 10% of the overall price of the receiver.

Specifications tell you what the box is supposed to do, but what the pragmatist really wants to know is whether it will work all the time, whenever it is needed. Reliability is what everybody dearly desires. "Radio, work. Don't just sit there and haunt me. Make me feel confident that you will be with me all the way. I'll tolerate a few idiosyncrasies now and then, as long as you are predictable and allow me to get the job done. And if you work as advertised all the time, you will be one of my proudest possessions."

It may be a small consolation, but reliability bugs the manufacturers perhaps even more than it frustrates pilots. Every manufacturer has a vested interest in producing reliable radios: boxes that do not work cost them money, lose them customers, and may raise the ugly specter of product liability. Very few manufacturers feel that there is an optimum relationship between quality-control costs and warranty costs. Most manufacturers believe that it is significantly cheaper to find failures before the unit leaves the shop than it is to discover them on a warranty claim, since the manufacturer's labor rate is about one-quarter to one-fifth the repair shops' rates and factory personnel can usually find the problem areas faster because of better component-test procedures and knowledge of the system. By means of extensive in-house testing the design engineers can develop a failure history and learn more about what can be done to prevent a failure in subsequent units. Returning the box to the factory on warranty isn't the answer — just ask any ex-giant killer like Radair. Manufacturers make their money turning out salable boxes, not repairing returned units for nothing.

Several avionics manufacturers have initiated programs to cut down on warranty claims. Narco has developed a specialized, computerized data-retrieval system to analyze problems discovered during the warranty period so that they can isolate trouble spots and correct them during production. All manufacturers have burn-in programs in which completed units are run under normal — and in some cases extreme — temperature-controlled conditions from 15 to 100 hours, depending on company philosophy. Such procedures are particularly effective in catching the infant-mortality problem so characteristic of solid-state components. Bendix believes that warranty costs determine whether or not profits are made in the avionics business, and they feel that the probability of bad boxes can be reduced if human intervention with the assembly process is minimized. Bendix industrial engineers recently installed relatively sophisticated labor-assistance machinery, such as an automatic instruction device programmed to direct a narrow beam of light on each assembly step, so that a worker can be led through all the intricate steps in the proper sequence needed to fabricate some of the more involved component sub-assembly boards.

Manufacturers go to considerable effort to assure reliability and control quality. The approach and extent of their individual efforts differ, but they all want to turn out units that will neither fail nor fall below usable operating performance. The number of people who work in the Reliability and Quality

Assurance Department at Collins is more than the total of all the employees at Genave, but differences run deeper than size alone. High-performance equipment is more sophisticated and complex, and in general the more components, the greater the probability of a failure within a given time period. To compensate for these risks, the designer uses higher-reliability components and, if possible, derates the performance that he demands from each unit or subassembly. Transistors and integrated circuits have a higher likelihood of failure during their first hours of use than later in their life cycle, but burn-in minimizes the problem to the customer. Frequency synthesizers are considerably more complicated than conventional crystal tuning techniques and are harder to repair if something goes wrong, so more inspections and tests are needed to guard against infant mortality. Navcoms such as Genave's Alpha 200 and Narco's Escort use simpler designs that utilize fewer and less sophisticated components, so reliability and quality-control procedures may differ in character — although not in purpose.

Only one avionics company out of six surveyed conducts random sampling of transistors and integrated circuits supplied by vendors. The rest use special test equipment to conduct incoming-parts inspections on 100% of their solid-state devices. Bendix, for example, uses a transistor- and diode-checker that is coupled with a computer to evaluate each solid-state component. Those that pass are heat-cycled and tested again. King also uses computerized test equipment to check completed subassemblies that contain integrated circuits. Collins has an extremely complete procedure for reliability and quality assurance. Each item that goes into a Collins avionics unit must be listed in a catalog of tested and approved components before an engineer is allowed to incorporate it in a design. Before a new Collins product is put on the market, 5,000 hours of laboratory testing are accumulated on the design, and the unit must demonstrate the desired mean time between failure (MTBF) initially specified as the goal for the design. (Failure means a box that doesn't work or performs below specifications.) Collins buys all its solid-state devices directly from the manufacturers and thus has a strong hold on the quality of component that they can expect. Other general-aviation-avionics companies generally buy from various sources and middlemen, depending on who can provide the best price and delivery schedule for the requested quantity. Collins uses metal-can, not plastic-encapsulated, transistors, and all components must meet or exceed mil specs, but even so, every one of the transistors and integrated circuits that Collins purchases is analyzed by a machine developed by the company called the Digital Pseudo Random Inspection Device.

In spite of exacting efforts such as these general-aviation avionics are plagued by a nerve-racking state of uneasy reliability. Most manufacturers admit that they expect 15% to 20% of their new products to produce a claim during a one-year warranty period. Ten-percent removal during the first year of use is a highly respectable failure rate by industry standards, and a few designers will admit that they do not feel that they have lost complete control until warranty returns reach the 30% removal level. Even with the comprehensive reliability and quality-assurance procedures used by Collins, the company's initial VHF-20 transceivers experienced field-use MTBFs that were about 20% of the target values demonstrated in lab tests. With data generated from actual operational use, however, a mature product emerged. Now Collins estimates MTBFs for the VHF-20 of about 1,800 hours. Accurate MTBFs are difficult to generate for general-aviation avionics because flight-hour records are generally not reported on warranty claims. A theoretical MTBF can be calculated from design considerations and known failure rates of its components, but laboratory-demonstrated MTBFs using production boxes are usually only one-third to one-half as long as the theoretical ideal, and actual field MTBFs are frequently from two to five times worse than even lab values due to poor installation and improper diagnosis of the avionics problem. John Ball, chief engineer of Narco Avionics, estimates that typical general-aviation MTBFs are approximately 4,000 hours for glideslopes and marker-beacon receivers, 600 hours for navcoms, 500 hours for transponders, and 300 to 400 hours for DME. As the units become more complicated, the reliability drops sharply. Based solely upon mature designs, normal reliability of good components and industry-accepted standards for locating problems during quality-control inspection and tests, the highest reliability possible during the first year of service is about seven removals, with 3.5 confirmed failures for every 100 units sold.

A frustrating lack of reliability unfortunately appears to be inevitable. With simpler designs, built-in redundant circuits, and innovative uses of solid-state components MTBFs will increase, but it is unrealistic to expect a piece of sophisticated avionics gear to match the reliability of a commercial FM-stereo receiver or a color-TV set. The established convention of using amplitude modulation (AM) for aircraft radios limits the technology transfer between avionics and commercial electronics, so significant reliability, performance, or cost enhancements developed in the home-entertainment field may not have a noticeable impact on avionics. Commercial electronics are extremely mature products compared with even the most popular aircraft radio. In one week of production at their Baltimore facility Bendix has produced more automobile radios than Narco sold Mark 12s during the entire history of that highly successful and popular navcom. Manufacturers are captives of the state of technology in avionics design and component selection, so, like it or not, we must learn to live with the situation as best we can.

Ah, but there are some things that a pragmatic pilot can do to fight back, provided that he knows what to expect,from his avionics gear, but more on that later. For now rely on Freud, who postulated that knowledge of your problems allows you to cope with them. Perhaps there is indeed an analogy between learning to live with avionics and remaining sane.

18.

MORE ABOUT AVIONICS

Reliability, the quality of consistently satisfactory performance, determines whether life with avionics will be bitter or sweet. Reliability is the characteristic most sought after when people buy avionics equipment, and, if realized, it can make the often frustrating marriage of necessity between pilots and navcoms a lasting honeymoon. But reliability can be elusive: if it is lacking, a pilot's life is filled with doubts, inconvenience, warranty claims, and potential trouble. What joy would abound if avionics were absolutely trustworthy! There would be no anxiety on dark nights or IFR. There would be none of the frustration and inconvenience that haunts pilots when an expensive DME refuses to lock on or an omni won't receive. If only a navcom with glideslope were as dependable as a VW — they cost about the same — then life would be sweet.

The day of trouble-free avionics is probably far off, but there are things that pilots can do to encourage dependability. Knowing the imperfections inherent in navcoms can save much anguish, and potential problems can thereby be avoided. We must live with the fact that radios fail with annoying frequency: try as they will, manufacturers can't seem to reduce warranty claims below 10% to 15%, and it is unlikely that major breakthroughs in circuit design or component technology will arrive soon to change that picture. Furthermore, if an avionics firm were to offer an advanced unit with redundant circuits embedded in large-scale metal-oxide semiconductor chips or some other concept that promises superreliability, it is unlikely that we would scrap all our existing gear overnight.

A measure of the reliability that a pilot can expect from a piece of avionics gear is the mean time between failure, or MTBF. It is simply the total time that a large number of identical radios have operated divided by the number of failures that the group experienced. MTBF is a useful figure, provided that you know what it really means. There are several ways to calculate MTBFs, depending upon what you want: the theoretical or maximum possible MTBF, the laboratory or demonstrated-under-ideal-conditions MTBF, or the actual MTBF derived from field use. It is the last that should interest pilots. A MTBF of 600 hours, for example, does not mean that a unit is certain to operate for 600 hours — quite the contrary. It means that the average time before failure for all navcoms of that type was 600 hours; some boxes failed sooner and some failed later. The probability that any particular unit will fail at some arbitrary number of hours is approximated by a simple reliability formula, which, unfortunately, does not account for the possibility of unusual failure patterns due to infant mortality or wear-out effect. The probability of a unit with a 600-hour MTBF having one or more failures is approximately one in six in its first 100 hours of use, two in five in its first 300 hours, and five in eight in its first 600 hours. The MTBF and the reliability formula state that there is always a finite chance that a navcom will fail, regardless of what its MTBF may be. A 1,200-hour unit obviously has a smaller chance of failure in 300 hours of use than the 600-hour box, and a 1,800 hour navcom has a still smaller failure probability. The likelihoods are about 22% and 16%, respectively, compared with about 40% for a MTBF of 600 hours. But a failure is possible, and, to cope with that fact of life, we must minimize the probability that the failure will jeopardize your flight.

The way to survive the failure-probability game is to use two or more radios. That may sound like asking for double trouble, but dual installations significantly reduce the chances of having a total avionics loss. The likelihood that a single navcom with a 600-hour MTBF will fail within 300 hours of use is about 2 1/2 times that of both units in a dual installation of the same radio going west during the same period. During a 1-hour flight the chance of both units failing is several hundred times less than the likelihood that a single unit will quit. On a price-versus-reliability basis radios with high MTBFs tend to be expensive, yet two lower-priced units with 600-hour MTBFs have about the same likelihood of failing within 300 hours as does one expensive box with a 1,800-hour MTBF. The less the time period compared with the MTBF, the more the odds favor a system of two inexpensive radios over a single expensive unit. Electrical-system failures, which are common to both units in a dual system, are obviously exceptions to the rule. Furthermore, dual installations consisting of high-MTBF gear offer significantly higher reliability than do dual installations of low-MTBF avionics.

For most general aviation avionics the concept of mean time between failures has more qualitative than quantitative significance. Except for rather high-priced equipment MTBFs are not available. Most manufacturers of lower-priced avionics do not conduct controlled tests to determine meaningful MTBFs, and flight hours between navcom failures for light aircraft are either not reported or not considered accurate. King Radio abandoned its efforts to establish MTBFs for its equipment because it found the data on confirmed failures to be unreliable. It is unfortunate that the avionics industry does not or cannot make a concerted effort to report realistic MTBF figures. For assessing reliability that number would be the most meaningful in a navcom's list of specifications.

Good reliability statistics are difficult to obtain because between one-third and one-half of all removals show up as non-failures when they are bench-checked. The difference between MTBF and MTBR (mean time between removal) depends largely on installation, operational, and maintenance factors that are beyond the manufacturer's control. Problems that cannot be related to a distinct failure in the offending unit are particularly frustrating, but it is precisely here that pilots can play an active role in improving apparent reliability.

Installation is critically important to achieving reliability in avionics. It is the glue that bonds together each element: if the installation is weak, the whole system comes apart. Voice communication is an example of this interdependency. Good results depend on a series of components that functions effectively: the microphone converts voice-pressure waves into electrical signals and usually cancels or reduces background noise; the communication unit modulates the selected carrier-frequency signal with the voice signal and produces the electrical energy that is converted into radio waves by the antenna; the aircraft's electrical system supplies the power used by the transmitter; and the various connectors and cables enable the mike, transmitter, antenna, and power source to work as a transmitting system. A poor microphone or an inefficient antenna installation can make a powerful transmitter less effective and seemingly less reliable than a properly installed low-power unit. A bad connector can render the most impressive radio impotent.

Antennas and the cables that connect them to transmitters are frequent causes of apparent reliability problems. A measure of antenna-system efficiency is the standing-wave ratio (SWR), which compares the power actually transmitted with the unused power that is reflected back to the transmitter. A perfect antenna has a SWR of one, which means that all the power that it receives is transmitted. A good broad-band antenna will have SWRs near one over the desired frequency spectrum, but a typical bent-whip antenna may have a SWR of six or more at some frequencies. A SWR of six means that about half the transmitter power is lost within the antenna system: a 5-watt transmitter would put out only 2.5 watts of real transmitting power. Good coaxial cable has a SWR of one, provided that the insulating material within the cable properly positions the center wire with respect to the metal outer conductor. If the cable is installed near heater ducts or directly under the roof of a plane that sits in the hot sun for long periods, the insulation, which usually is polyethylene, can melt and cause the center wire to shift. That results in higher SWRs and serious degradation in transmission performance before the cable shorts out completely and the system fails. Coaxial cable with more heat-resistant insulation materials such as Teflon is available, but it is too costly to use. A good avionics shop has the equipment to check the standing-wave ratio of the antenna/cable combination. If the system checks out well for transmission, it will also function properly for reception.

Heat is the source of many installation-related problems. In spite of what the ads say, the new transistor navcom units can be affected by heat. It is true that they require less cooling than the older vacuum-tube radios, but heat starts to degrade the performance of power transistors at about 25 C and can cause the device to fail at values above approximately 125 C. The corresponding maximum temperature for a vacuum-tube unit is about 85 C. The installation often accounts for the very high temperatures that can trouble panel-mounted radios stacked one on top of another.

Water is another installation-related problem producer. If water gets into the heart of a radio, your expensive navcom may never be the same. Leaky windshields are the source of considerable trouble for panel-mounted units, and leaks from poor beacon or antenna installations can plague remotely mounted avionics. Beware of nose-well installations for remote units: that area is particularly susceptible to water, dirt, and other foreign objects that help generate reliability problems. Connectors especially are vulnerable to moisture. If these key links in the system become corroded or damaged, even the most expensive and supposedly reliable navcom will be brought to its knees. Many frustrating avionics troubles such as garbled audio reception and intermittent cutouts are frequently caused by a short in a seemingly insignificant connector.

To extract maximum operational utility from your system, start by knowing what to expect from each unit. Consistent satisfactory performance is the meaningful quality to look for. For example, the ability of a navcom to transmit and receive signals depends upon radiation patterns that emanate from and are received by the antenna. Depending on where it is located on the aircraft, a particular antenna may experience interference from the fuselage, the wings, or other antennas. It may consequently be possible to communicate perfectly with a station ahead and slightly to one side of the aircraft but impossible or at best marginal to contact a second radio facility located in another direction. No need to panic and assume that the radio just failed − instead note the orientation of the silent station, then try turning the aircraft and observing whether the reception changes. Chances are that it will. Problems with distorted antenna patterns are often associated with the frequency you've tuned on the navcom as well as the frequency tuned into other units in the aircraft. Look for such patterns and recognize them as characteristics of the particular aircraft installation, not as indications of an avionics failure. Also keep in mind the significance of proper antenna location and seek the advice of a competent radioman when additional avionics are installed.

Navcom failures are not always sudden and complete malfunctions. A unit's performance may deteriorate over a period of time, but the decision to remove the box for repairs should depend upon the ability of the radio to produce satisfactory results: transmitter range may have diminished or the aircraft may have to be near the LOM for the ADF to give a good relative bearing. If the reduced performance is sufficient to get the job done and the pilot can depend upon receiving consistently satisfactory results once he is within range, the unit is still reliable and can give additional hours of useful service. One advantage of high-power transmitters is their ability to lose some of their original power and still produce satisfactory performance.

The ability of the microphone to function properly has a lot to do with good transmitter performance. If the microphone gain is too high, over-modulation occurs and transmissions are muffled and distorted. On the other hand, if the gain is too low, transmissions are weak. To check whether the mike gain is matched properly with the transmitter, have a friend listen to your transmissions or listen to your own sidetone. Your voice should sound about as good as it does over the telephone. Listen to the signal transmissions with the microphone placed in the normal position for communicating but do not talk. The noise level should not be too high. If your transmitter fails these tests, have a good radio shop check the modulator level and adjust it to match the mike gain. Many of the new mikes have a wide range of adjustment outputs, so this check is very important, particularly if you install a new mike or transmitter.

Noise-canceling microphones, if properly used, can increase the effectiveness of your transmitter. They have two openings where sound enters the mike. One is the opening that you talk into; the other is usually on the back or bottom. Sound entering both openings, such as noise, is canceled, but sound that enters just one opening, such as your voice, is picked up by the microphone. If you hold the mike away from your lips so that your voice enters both openings, your voice will be canceled just as if it were noise. For proper operation hold the mike very close to your lips. To check the effectiveness of a noise-canceling microphone while in flight, hold the mike away from you and key it. Listen to the sidetone with your headset: excessive background noise should be canceled. Place your fingers over the noise-canceling opening of the mike, and you'll find out how effective the noise-canceling feature is.

Boom microphones usually pick up a lot of background noise because they are positioned away from the mouth, and their sensitivity is adjusted quite high in order to pick up voice. Unfortunately, noise is also picked up. By careful adjustment of the modulation level and by positioning the mike close to the lips, noise pickup can be reduced substantially. Some boom mikes can be obtained with noise-canceling elements.

Even the standard carbon microphone can be a source of transmitter troubles, because the carbon grains that turn the voice vibrations into electrical signals tend to become packed together with age and cause poor performance. New elements can be obtained for considerably less money than a new mike, but if they are not available, you can fix a carbon microphone with a good whack against something hard. That loosens up the granules and improves performance. Before you try this, make sure that you have a carbon mike. Many of the units sold today are the transistorized dynamic type, and giving them a whack will just add to your problems.

Dual-microphone installations can cause problems if the mike jacks are connected in parallel, an arrangement typical in general-aviation aircraft, because any short that occurs in one mike will short out the entire circuit, rendering both mikes useless. Transistorized dynamic mikes are more susceptible to internal shorts than carbon types. To troubleshoot this problem, pull both mikes out, then plug one in at a time to see if the transmitter will work with at least one mike. Be sure that the mike plugs are fully seated in the jacks, since a plug that has partially come out can cause a short and make both mikes inoperative. Beware of pulling mike leads out of jacks by yanking on the cord instead of taking hold of the plug case: that can break wires in the cable. Dual-microphone installations are still worthwhile because of the redundancy that they offer. An otherwise reliable avionics system is useless if your one and only microphone gives up the ghost.

Because the quality of the average speaker used in general aviation avionics is poor, you'll hear a lot better by using headphones, which cut out cockpit noise and give better audio reception. Background interference can be reduced further without diminishing audio performance by using earplugs. To illustrate this the next time you fly, try listening to the speaker while your fingers are blocking your ears. You'll find that the audio is actually more distinguishable because you have blocked air and engine noise. Another pressing reason to use earplugs is that headsets do not filter out all the low-frequency sounds that are most damaging to your ears. To preserve your hearing and improve your audio reception, wear a set of earplugs under a headset and adjust the volume accordingly. You will be amazed at how your audio reception improves.

Another feature that can improve the overall performance of a navcom is an isolation amplifier and a switching panel, a combination that separates the audio input from each radio, amplifies the selected input, and directs it to the speaker or headphones. With such a setup, a malfunction of the input from one set will not affect the audio available from other units; more significantly, the iso amp eliminates audio-loading problems and the distortion that can result when too many sets are connected in parallel to one speaker. Without an iso amp the audio power of each set tries to drive both the speaker and the other radios, so the audio that we hear is reduced. A switching panel by itself does not eliminate the loading problem: an isolation amplifier is required to eliminate the possibility of weak audio when a number of navcoms are connected to the same audio system. Many radios available today have an abundance of audio power, so loading is not a problem in unsophisticated installations, but the combination of lots of avionics and weak audio indicates the need for an iso amp in addition to a switching panel.

The two most effective actions that a pilot can take if avionics act up are to select a good radio shop for troubleshooting and maintenance and to communicate your needs accurately. The latter is particularly important. Charles Husick, president of Narco Avionics, likens the task of a radio repairman to that of a pediatrician who must determine what is wrong with a one-month-old baby. At least the infant can scream when the doctor pokes and probes; a radio just sits there. It is up to the pilot to describe exactly what went wrong and under precisely what circumstances. Tell the repair shop the flight conditions (location, altitude and OAT), the aircraft configuration (manifold pressure and rpm, gear up or down), the status of other avionics equipment (including exact frequencies of all units at the time of the problem), and the operating state of the troublesome unit. If you are lucky, the bug might be a gear door that shields a transponder or DME antenna or something simple like that. Give a history of the radio's performance characteristics and include a list of any changes made recently to the airframe. Has an ELT been installed? Was the alternator worked on? Was the aircraft stored outside instead of in its usual place in a heated hangar? Has the plane stood for a long time without the avionics gear being operated? Frequent use of navcoms, even on the ground, keeps out moisture and prolongs useful life. The details will help the radio shop and save you money and frustration.

Before you pull an apparently bad radio, look at the avionics installation. Operate the radios on the ground with the engine shut down and with the engine running to determine whether the electrical system is causing the trouble. If your problem is audio noise, turn on all the avionics gear unit by unit and observe whether the noise gets worse; if it does, the source of the trouble is probably the alternator. Noise can also be caused by a bad battery. If you have a dual installation of identical units, switch them and observe whether the problem moves with the box or is confined to a particular location. Removing a reliable radio with a poor or defective installation from an aircraft runs up bills and tempers and exposes your navcoms to the hazards of unnecessary probing, which may create real problems.

Selecting a good radio shop is more difficult than describing what went wrong. Call the navcom manufacturer and ask whom they recommend in your area. None of the firms is too big or too small to offer assistance, and they welcome the opportunity to direct you to shops that can handle their equipment without fouling up the unit further. Also look for radio shops that have a clean, neat appearance and have manuals that apply to your particular gear and are up-to-date. Observe which types of aircraft are in for service and ask other customers whether the service is satisfactory. When you locate a man who knows your equipment and can handle your needs, be nice to him. He could be the best friend that you and your radio have.

Good radio maintenance is hard to find, so it pays to choose a reliable, trouble-free unit and to treat it correctly from the first day that you turn it on. A navcom brand and type that has been around and selling for several years is likely to have fewer problems because it is a more mature product, but the degree of maturity is only relative. The unit still reflects the overall climate of general aviation avionics. A unit with a weakness will have a poor field reputation; you can discern that by talking to several reliable radio shops and aircraft operators. Look for real features, not just spectacular specifications — complexity can cause problems and make repairs difficult. Once a unit has undergone an extensive burn-in, the likelihood of failures in the solid-state components is reduced, so look for a company that uses this procedure in the manufacturing process. Mechanical components do not benefit from burn-in, unfortunately, and they remain a potential source of trouble. Look for good mechanical design in knobs, on/off switches, and all moving parts. Once it is past the infant-mortality stage, frequency synthesis offers more reliability than mechanical tuning techniques, but the synthesizers are too complicated for many shops to troubleshoot. Avionics that are packaged neatly, with the printed circuit boards carefully laid out and readily accessible, offer a certain degree of maintenance ease when the seemingly inevitable failure occurs. To put off that day as long as possible, read the owner's manual and abide by it. Learning to live with avionics requires patience, understanding, and a certain degree of pragmatism. At times the task can be sheer frustration, but those curiously fascinating boxes open up highways through the most forbidding skies. They link us with civilization as we venture outside it. As a pilot you may find that to tolerate the shortcomings of avionics is a small price for such rewards.

19.

AUTOPILOTS

Most of us would rather drive than be driven and fly rather than be flown. Whether it is because each person feels himself superior to all others or because of tiny differences in reaction time or piloting technique among individuals, most of us feel uncomfortable during someone else's takeoff or approach. We have to wrestle with the temptation to backseat-drive or even to give the yoke or throttle a surreptitious nudge from time to time.

Imagine, then, the task that faces the maker of an autopilot. He must create a device that will more or less satisfy all pilots that it is doing as good a job as they would, that it is free of annoying habits that would make its competence suspect, and that in addition it can do a job that takes most humans years to learn to do well. Furthermore, even if it is malfunctioning, it must be reliable, serviceable, and free of murderous tendencies.

Other pieces of airborne equipment face rigorous requirements, but the autopilot is unique in that it is designed to behave like a human pilot — at his best. Radios process electromagnetic radiation and other instruments measure things that we can only vaguely sense, but an autopilot flies as you do, and from the outside looking in it is hard to tell the difference in normal maneuvers between a human pilot and an artificial one. That is the autopilot's great trick, and achieving it involves an artistry that goes beyond mere meshings of hardware and maintenance-manual math.

Autopilots cover a wide range of performance. The most basic, the wing-levelers, simply augment stability in the roll axis, which is the axis of lowest inherent stability in aircraft. So simple is this function that most pilots would hardly apply the title of autopilot to a wing-leveler, especially since more complex autopilots can perform navigational tasks that range from holding a heading to executing a complete flight, with the pilot's intervention consisting solely of choosing the destination and telling the black box when to start.

It is possible in talking about electronic or mechanical devices to speak of their "knowing" this or that about what is going on around or within them. You might say that an auto engine "knows" how fast it is turning over - the speed is registered in stress levels in the moving parts, cooling rates, and so on — but does not "know" which way it is headed. A landing-gear sequencing system "knows" its status by "interrogating" various limit and control switches. Such language may seem fancifully anthropomorphic, but it is really hard to see in what way the autopilot's "knowing" that the plane is in a bank differs from *your* knowing that it is. Turning things around, a nonhuman observer watching a pilot at work and responding to what it "senses" and "knows" about what the airplane is doing would note that the human acted like a highly complicated automatic-control system but would infer nothing about what we call "consciousness." In fact, how much a "living" flier is like an automated one can be seen in the fact that grasshopper flight is controlled by feedback and servo systems that are no more complicated than those in a three-axis autopilot. A man can fly an approach to a landing; so can an autopilot, and it does not seem to need an "immortal soul" to do so.

Human pilots get position information from visual cues and to some extent from the balance mechanism of the inner ear, though the latter is misleading if visual cues are absent. Autopilots have in their gyros an attitude reference that is independent of visual cues and capable of holding a fixed position in space over a long period of time. An ideal gyroscope retains its position with respect to the heavens so that, once it is set in motion, it maintains its alignment independently of the rotation of the earth. To make a gyro refer to the center of the earth rather than to the stars for its sense of up and down, an erected mechanism is supplied, which also corrects small errors introduced by friction and by the inevitable imperfections in the gyro's manufacture.

The information contained in the gyro's position is translated into power, which the autopilot amplifies to drive the airplane's controls. Early autopilots used air pressure supplied to movable diaphragms through ports the opening of which was a function of gyro position. Their movement operated valves that supplied oil under pressure to hydraulic servos. Some systems use a vacuum instead of hydraulic pressure, but the controlling principles are similar. The majority of contemporary autopilots, however, use electrical power.

The gyro has two measurable properties of which the autopilot is aware: position and rate of change of position. Position information enables the autopilot to intercept and hold a preset heading or altitude, while rate information permits it to control the speed of its own corrections, adjusting them to the situation at hand. Position and rate information can also come to the autopilot from radios or from the pilot's panel controls, but the short-term stability of the device depends on its gyros. As a man threading his way through a forest remains upright and walks forward by motor instinct and navigates more or less through his brain, the autopilot keeps the airplane upright solely by means of its gyros and guides it both by the gyros and by various other means.

This sounds straightforward enough until you reflect on the series of cues and responses by which a human pilot makes, say, a change of heading. First he decides that he wants to turn to another heading; perhaps he is changing course or perhaps returning to a heading after being blown off-course by a gust. He turns the yoke in the direction he wants the airplane to turn, and the ailerons deflect accordingly. He proportions the yoke movement, like a driver rounding a curve, to the response that he feels in the vehicle. If he tries to make a right turn and if at the same moment there is a strong lifting eddy under the right wing, he will feel no roll response to the normal yoke movement and will tilt the yoke farther than usual. When the desired rate of roll is attained, the pilot neutralizes the desired angle of bank to arrive. He proportions the bank angle to the desired heading change. As the desired bank angle approaches, the pilot turns the yoke away from the turn to arrest the rolling movement; he again proportions his yoke movement to the rate of response of the airplane, and he again neutralizes the controls when the rolling movement stops. As the desired heading approaches, the pilot reverses the entire procedure, adjusting his final yoke movements so that the airplane not only rolls out level but on heading. The difficulty in simulating this procedure mechanically lies in the nonlinear relationship between aileron position and airplane movement or angle: the airplane may be rolling to the right at the same time as the yoke is angled to the left.

The pilot decides he wants a new heading by consulting the directional gyro; the autopilot does the same, comparing the current position and rate of the DG with instructions that it has received from the human pilot. If the actual heading and the desired heading coincide, the autopilot, like the human pilot, feels no desire to do anything; but if the proper heading is different from the real one, the autopilot feels a voltage the polarity of which depends on the direction of the heading error. Voltage is the same as desire. At this point, the human pilot may or may not take the unnecessary step of "being aware" of the need for correction; the autopilot bypasses this step and channels the voltage via an amplifier to a servomotor that controls the ailerons. A servomotor is a variable-speed, reversible electric motor with a huge down-gearing ratio − perhaps 10,000:1 − and equipped with a mechanism that permits it to decouple itself from the control system and with a friction clutch that limits its authority. In response to the amplifier voltage the servomotor spins up and the ailerons move.

At this point what is called a follow-up enters the picture: without it the ailerons would go on moving to their stops, since the voltage created by the heading error is still present. In order to stop the ailerons, another voltage − equal in magnitude and opposite in sign − is supplied by the ailerons themselves. This was originally done by means of various feedback systems, which informed the autopilot of the control's position. It can now be done by an electronic synthesizer, built into the autopilot circuitry, that exactly mimics the response of the ailerons to motor input voltage. As the ailerons are deflected, they produce more and more voltage, which eventually balances the DG voltage out to zero. Since the movement of the servomotor is a function of amplifier voltage, when the aileron reaches a certain deflection, it stops moving. (The deflection is proportional to the heading error up to an arbitrarily imposed limit of perhaps 30 degrees of bank.)

In the meantime the airplane has begun to roll, and its heading to change. The gyros are now supplying these two new facts to the amplifier in the form of voltages that are opposite in sign to the original heading-error voltage; the original voltage is not only balanced but overbalanced by the aileron-position voltage. There is now an opposite voltage going to the servomotor, which in response reverses its direction and moves the ailerons against the turn in order to stop the roll. By tuning this system in such a way as to accommodate the voltage threshold of the motor and whatever other lags may appear, it is possible to get smooth roll entry and halt and to hold a steady angle of bank in the turn.

As the actual heading approaches within some number of degrees of the desired heading − the number is selected by the autopilot manufacturer and may be around 20 degrees − the voltage generated by the heading error begins to drop, aiming at zero, at which the two headings coincide. Now the voltage coming from the bank angle and the steady azimuth change exceeds the heading error voltage, and the servomotor gets a gradually increasing signal to deflect the ailerons out-of-turn. The aileron follow-up again comes into play to limit the rate of roll-out, and the rate stabilizes: when the wings approach level and the heading error approaches zero, the only remaining voltages of any significance are those arising from rolling motion and nonlevel wing position, and both are opposite in polarity to the direction of the rolling movement. The servomotor therefore initiates a stop-roll aileron deflection; the follow-up once again synchronizes the deflection with the airplane's response; and the end result is that all voltages become zero when the airplane is level, on heading, with zero roll rate − that is, it has satisfied and voided all its "desires."

In addition to this mechanism, which responds to position inputs, there are *rate networks*, which measure the velocity of rolling, pitching, or turning motions and modify the autopilot's responses accordingly, limiting its own rate of roll to a comfortable value and making its response proportional to the input. A wing slowly dropped in level flight is slowly raised, while a sudden disturbance by a gust is more rapidly corrected. Other circuits compensate for drift of the zero point (when one wing contains more fuel than the other, for instance), permit manual trimming, and, in some wing-levelers, store electrical energy generated during turning disturbances to be discharged during the correcting maneuver so that the autopilot overcorrects and not only restores level flight but brings the airplane more or less back to its original heading.

Once the mechanism by which the autopilot deals with roll and heading change is understood, it is not hard to visualize how it is applied to the pitch axis. There are differences − for example, the airplane is stable in pitch, and it is unnecessary to neutralize the elevator after initiating a pitching maneuver − but the basic principle of voltage imbalance leading to correction that eventually produces a new balance, all mediated through a follow-up, still holds. In pitch, there is the additional requirement of an automatic *trim* system, which, after determining that there is a load on an elevator actuator, moves to eliminate that load. The autopilot wants to maintain a programmed pitch angle, which is analogous to the level-flight position in the roll axis, and maintains a preset altitude, which is analogous to a selected heading.

Since the autopilot's knowledge of the position and attitude of the airplane consists of voltages generated by gyros, it is a simple matter to feed commands into the system by means of potentiometers that can be adjusted by the human pilot to simulate a gyro voltage. To command a right turn, you manually present the autopilot with the same voltage as would a gyro that indicated that the airplane was making a left turn. Heading selectors, altitude selectors, VOR, localizer trackers, and so on are all sources of voltage that the autopilot tries to nullify by maneuvering the airplane in such a way as to produce opposite voltages.

The single-axis or wing-leveler autopilots are ignorant of the airplane's position or heading, and concentrate on maintaining a certain attitude about the roll and yaw axes. Some can turn the airplane by input from a knob and be trimmed to hold a heading by another knob, which controls a small voltage intended to balance out uneven fuel loads or electronic biases within the circuitry. For a little more money they can be made to track an omni radial inbound or outbound: the voltage that swings the needle is used to simulate a gyro input at the amplifier, and the airplane turns toward the needle just as it would turn out of an inadvertent bank or toward a newly selected heading.

One of the most difficult problems confronting an autopilot manufacturer is the *interface* between his autopilot and the avionics that will be used with it. The autopilot manufacturers supply their own gyros and can tune the system to the gyros as nicely as they wish, but the avionics are usually supplied by the pilot, and the autopilot has to work with any radio equipment. This may sound simple, since all radio equipment seems to be doing the same job, but in fact different radios use different methods to obtain the final result − the panel display of the omni head, for example. Cheaper radios do a less delicate job of signal discrimination than do more expensive ones and make the inferior signal suffice by damping the display to give it an artificial steadiness. The autopilot normally wants to work with the raw signal, but if it is of poor quality, as when an inexpensive omni receiver is working a station 50 miles away, fluctuations occur that are so large that the autopilot is unable to hold a steady track. The autopilot in effect feels that it is tracking a wide or blurred needle and is constantly in doubt about which part of the needle to aim for. It has become common practice in inexpensive installations to couple the autopilot to the omni needle rather than to the raw signal; the needle is heavily damped − the more economical the set, the heavier the damping is likely to be − and provides the autopilot with an averaged signal, which it can track steadily.

The variety of tricks that an autopilot can perform is endless, but, once the basic control mechanisms exist, "complicated" navigational procedures can be programmed into the autopilot rather easily, if expensively. Automatic navigation is a cut-and-dried affair − a three dimensional flow chart laid out against the sky. The elegant ingenuity of autopilot design is to be found mostly in the basic piloting mechanism.

For economic reasons few autopilots used in light aircraft are capable of extended automatic maneuvering. They typically perform in "modes" that are selected by the pilot by pushing buttons or turning selector switches. Allowing the pilot to lead the autopilot carefully through an extensive maneuver such as an approach simplifies the logic circuitry and therefore lowers the cost. It may also be advantageous in that the pilot is forced to remain abreast of events, since no step will be taken by the autopilot until the necessary preceding step has been completed. For example, an ILS approach is normally performed by a middlebrow autopilot as a series of discrete intercepts, turns to selected headings, lock-ons, holds, and so on, with each step selected and monitored by the human pilot. The great advantage of working through the autopilot is that the job of the human pilot becomes administrative and therefore intermittent. The airplane is kept upright and stable by the autopilot. Either because the reliability of autopilots is regarded as at least equal to that of humans or because pilots' tastes for gadgetry know no bounds, increasingly expensive autopilots can be persuaded to remove more and more elements from the pilot's role at a cost of many thousands of dollars per button-push eliminated.

In spite of this upsurge of confidence even simple autopilots have a lingering reputation of unreliability. This may be because most autopilots were built at least a few years ago. It may also be the result of a generally low priority given to autopilots by owners, renters, and repair shops. Radios get fixed first, so that is where the money, experience, and concerns on the part of repairmen lie. Their training is not uniformly high in quality, and there is a tendency to make additional money by not making a total repair the first time. In fact, it is hard enough to find somebody who will do a lasting repair on a radio, much less on an autopilot. Still autopilot reliability has improved considerably in recent years, and the quality and exotic capabilities that were inaccessible a few years ago are now within the reach of most manufacturers.

If autopilots are indeed becoming more reliable and sophisticated, so much the better, because they are also becoming more indispensable with every passing year and notice of proposed rulemaking. It may have been possible to hand-fly a solid IFR trip in the lost days when traffic was light, the route structure was simple, and half the time you weren't even in contact with ATC. Today, while the en-route segment has been made uneventful to the point of being tedious − thanks to the introduction of transponders, DME, and better radar − departure and approach procedures in unfamiliar terminal areas can be difficult for a single pilot of average ability to hand-fly. An autopilot is vital, and even a pilot who flies only infrequent light IFR can hardly help being seduced by the alleviation of strain that a mere single-axis with omni coupler can provide. Such an autopilot costs about $1,000 and can give a convincing imitation of a $2,000 two-axis unit.

Capable as autopilots are, pilots often expect too much of them. After all, their controls look simple, and people imagine that it is not even necessary to read the operating manual in order to get the most out of them. Consequently, the devices are expected to do the impossible, while their real capabilities, which often exceed expectations, go overlooked. On first meeting one, it is a good idea to read the handbook through a few times to make sure that you understand what it says. This is not always easy, since firms that spent millions developing an electronic device sometimes seem unaware that a few extra dollars would be well-spent on hiring a professional writer who could make their product usable. If you can, get a factory representative or an instructor who is familiar with the autopilot to give you a demonstration of its use, for, even if you have read the book thoroughly, you will still have a few tricks to learn.

After a few days, weeks, or months of use something is almost bound to go wrong − such as endless roll, nutation, or snaking. Rather than succumbing to the temptation to turn the autopilot off in disgust and take it to a shop because "it doesn't work," try to describe the exact nature of the trouble, including the conditions under which it occurs, when it first occurred, and what it results in. Specific malfunctions often point to specific circuits. If you have any doubts about the competence or experience of the shop involved, call the manufacturer's headquarters and find out about their experience with that shop, if any, or their recommendation of another in the area. For electronic gear factory service is usually best. It is faster and cheaper than shop service and as good or better. A factory can usually provide one-day service, but it prefers to see the autopilot together with the airplane rather than disconnected pieces that you might remove and send. If you have chronic trouble, a visit to the factory is perhaps the only solution.

Despite their relatively high price − with a couple of exceptions all autopilots cost more than most navcoms − autopilots will become more and more common in the coming years. The complexity of the IFR system makes them welcome. A properly adjusted autopilot can be expected to run reliably − more reliably, in fact, than most radios, because it has fewer parts. All pilots will probably become as familiar with autopilots' capabilities and limitations as they now are with those of airplanes and engines. The dubious possibilities of completely artificial stabilization and command navigation for light aircraft are not so remote as they might seem. Make friends with the next autopilot you meet: it might turn out to be one of the family. At any rate his resemblance to you is uncanny.

20.

FLIGHT DIRECTORS

Flight directors have been available for about 20 years, yet there are still far too few pilots who really understand what they are, how they work, and why they are valuable. Manufacturers of the devices still receive letters referring to horizontal-situation displays as flight directors although this useful instrument, which superimposes heading and navigational presentations, is only a relatively small portion of a flight-director package. Pilots frequently ask why they should spend the additional money for a flight director if they have an autopilot. Some people simply equate flight directors with large aircraft and ignore the issue when it comes to sophisticated singles and light twins.

A flight director gives the pilot a visual display that tells him *what* to do with the aircraft attitude and *when* to do it. To intercept an omni radial without a flight director, for example, a pilot observes the position and motion of his omni needle, figures out the angle at which he is intercepting the on-course, and decides when he should start his turn and what angle of bank he should use in order to be heading in the proper direction exactly when the omni needle centers. A flight director considers the same factors for him and automatically computes when the aircraft should be turned to intercept the radial; it also tells the pilot how much the plane's attitude should be changed to accomplish the task. The pilot is thus spared the work load of mentally adding up information from several instruments and figuring out how much bank attitude he should use for his turn and when he should start the maneuver. Furthermore, a flight director performs these computations with the precision of a pro and with twice the speed.

A flight director imparts this "command" information by means of a display superimposed on the aircraft's gyro horizon. The display can take the form of ILS-type cross-pointers, as in the Sperry Stars IVB; a single V bar, as in the Edo-Aire/Mitchell Century IVFD; or two wing-tip command dots such as those in the Bendix FDS-840. Although these three methods of presenting flight-director information are known by various trade names and have their own vociferous exponents and detractors, they all function in essentially the same manner: they show the pilot what the aircraft's attitude should be to intercept and maintain a desired flight path.

The Sperry Stars IVB system of presenting guidance information falls into the class known as *dual cue*. Pitch commands are made by a horizontal glideslopelike cross-pointer, and heading or bank commands are given by a vertical cross-pointer. When attempting to intercept a chosen omni radial, say, the flight director computes the necessary bank angle and moves the vertical cross-pointer in the direction of the required bank. When the pilot notes the displacement of that heading cue, he starts a turn in the indicated direction. As he does so, the vertical cross-pointer starts to center, and, when it is centered, the pilot has achieved the correct bank angle and thus the rate of turn needed to intercept the omni radio. As the plane nears the on-course, the vertical cross-pointer moves in a direction opposite to the way in which the plane is turning, indicating to the pilot that he should adjust his bank angle by turning toward the cross-pointer. The needle recenters when he has achieved the correct bank angle to satisfy the commands of the flight director. By turning toward the vertical cross-pointer in a manner that keeps the needle always centered on the face of the flight director's gyro horizon, the pilot can fly an attitude that will intercept and track any desired heading, omni radial, or localizer programmed into the flight-director controller. The horizontal glideslopelike cross-pointer of a dual-cue flight director provides pitch-attitude commands in a manner somewhat analogous to the way in which the vertical cross-pointer presents bank-angle guidance. The pilot programs the director to intercept and track a particular rate of climb or descent, pitch attitude or glideslope. By lining up the aircraft's position, as shown on the gyro horizon, with the flight-director command cross-pointer, the pilot constantly flies at the proper attitude to intercept and stay on the vertical-mode requirements that he wants.

Single-cue flight directors operate in a manner similar to that of their dual-cue cousins, except that both horizontal- and vertical-mode guidance are presented concurrently on the same command device. In the Edo-Aire/Mitchell Century IVFD, the King KFC 300, and several Collins models, such as the FD-108, FD-109, and FD-112 V, pitch and bank commands are presented by a V bar that pitches and banks to show the pilot how he should adjust the aircraft's attitude. Pitch commands are identical to those of the dual-cue glideslopelike cross-pointer, but bank commands are presented by a simple banked-command display instead of the lateral displacement of a cross-pointer. The pilot adjusts the plane's attitude so that the "airplane" in the flight-director display snuggles up against the V bar, which produces the proper attitude for the maneuver.

The command-dot type of presentation found on the Bendix FDS 840 and the ARC/Cessna 300 IFCS, 400 IFCS, and 800B IFCS falls within the general category of single-cue flight directors, because bank and pitch commands are presented on the same device. The desired attitude is achieved by placing the wing tips of the gyro horizon's reference aircraft on small dots or paddles that move collectively up or down to command pitch changes and differentially up or down to give bank guidance. By keeping the wing tips and the dots aligned, the pilot achieves the proper pitch and bank attitude.

Whether a pilot prefers a single-cue or dual-cue presentation is a matter of personal choice. Both systems are easy to fly, and several manufacturers offer both single- and dual-cue versions of the same flight director. The dual-cue Sperry Stars IVB and the single-cue Stars IIB is an example; the dual-cue Collins FD-112 C and the single-cue FD-112 V is another. The V-bar type of single-cue presentation is slightly more complicated internally than either the cross-pointer or command-dot types, so it costs more. Some pilots prefer this type, however, because the bank commands are straightforward and natural. Some airlines prefer the dual-cue presentation because on very low approaches, they like to see lateral displacement rather than bank angle on the heading command needle; airline pilots are reluctant to do too much banking at low altitudes with their long-winged aircraft, particularly if the engines are hung below those wings.

The command presentation shown to the pilot on the gyro horizon − or ADI (attitude/director indicator), as it is called in flight-director parlance − is the heart of what the pilot sees and uses when he flies with a flight director. There are other very important components, however, and together they constitute a flight-director system. One of these, the HSI (horizontal-situation indicator), is always included with the director package, although the instrument does not present computed guidance information such as the ADI does. This is not to imply that the HSI is not a useful and effort-saving device. It is, but its utility derives from its ability to combine heading information and omni or ILS information in a mechanical way, allowing the pilot to see the position of his plane in relation to the navigational situation regardless of whether he is heading to or from the VOR or ILS transmitter. The HSI also presents raw glideslope data by means of an indicator located along one side of the instrument.

The HSI is so useful that units have been sold separately − not as part of a flight-director package − quite successfully for several years now. The Collins PN-101 is the granddaddy of the HSI set. This useful and time-proven instrument has simplified VOR/ILS orientation and tracking problems for years with its easy-to-interpret pictorial view of the heading and navigation situation. The Narco DGO-10 also offers the general aviation pilot a bird's-eye picture of the aircraft heading in relation to the on-course.

The key words in the description of the ADI and HSI are "director" and "situation." The attitude/*director* indicator displays the computed commands which tell the pilot what to do and when to do it in order to achieve the desired results. The horizontal *situation* indicator presents raw or noncomputed information, which depicts the actual situation as it exists at the moment. Thus a pilot can satisfy the ADI commands but not be on course. He will be intercepting the on-course at the optimum angle, but only his horizontal situation indicator will show him exactly where he is with respect to the desired course.

Some people criticize flight directors because it is possible to wind up with the desired situation by concentrating solely on the command bars. They claim that a pilot can abuse this feature and "tunnel in" on the ADI at the expense of obtaining an exact picture of his total situation. This criticism is somewhat weak, however, since the problem lies in the failure of the pilot to set up an effective cross-check pattern rather than in a shortcoming of the flight director.

Depending on the degree of ADI sophistication, raw or situation information is also presented on the attitude/director indicator. The Collins FD-112 series offers raw glideslope readings on the ADI but not on the HSI. Sperry Stars flight directors present glideslope displacement on both the ADI and the HSI. The Sperry units also present an expanded localizer displacement and an altitude indication from about the middle marker to touchdown. The Collins FD-108 and FD-109 series also offer these features, which are used to check the relative position of the aircraft with respect to the runway centerline but are too sensitive to be used for tracking. The ultimate in situation information presented on the ADI is found in the ultrasophisticated Bendix FGS-70. This instrument presents raw localizer and glideslope data in the form of a bull's-eye located in the center of the attitude/director indicator. The fixed bull's-eye cross-pointer moves both up and down and right and left to show the plane's position relative to the ILS centerline. Auxiliary-situation references provide radar attitude and fast-slow angle-of-attack information. The Bendix FGS-70 ADI situation information is for reference only: the pilot flies the single-cue command bars for guidance.

The brain of a flight director is a computer that functions in a manner nearly identical to an autopilot flight computer: in some cases it is the same computer. This device receives instructions from the pilot — via pushbuttons — for particular flight modes (such as intercepting and tracking an ILS) and factors them with the aircraft's heading and closure velocity relative to the navigational signals that it is receiving from the VOR or ILS receiver. The pilot's instructions are set up on the flight-director controller, which also serves as the autopilot controller on models, such as the Edo-Aire/Mitchell and the Bendix FCS-810, that are sold primarily as autopilot/flight-director combinations. The other signals come from gyros and the primary navigational radio, just as they do in an autopilot system.

The computer compares what the pilot has instructed it to do with what the plane is actually doing. It computes the difference, or error signal, between the desired and the actual situations and uses this error signal along with other inputs to position the flight-director command bars. These signals may even be the same as those used to control the position of the autopilot servos; in all cases, they are similar. With a flight director, the pilot uses his muscles rather than the autopilot servos to move the aircraft's flight controls. As he does so, the computer senses whether the aircraft is responding satisfactorily and makes the appropriate adjustments to the command bars in the same way that an autopilot's computer controls the servos.

In essence, the only difference between the internal working of a flight director and of an autopilot is that in a flight director, the pilot satisfies the error commands of the computer; in an autopilot, feedback position servos do the satisfying. In other words, the pilot closes the control loop in one case and the position servos close the loop in the other. The computer and data sensors can be identical.

The pilot's being in the control loop is one of the major advantages of a flight director. Research has shown that pilots can detect and correct malfunctions faster if they are part of the system than if they are merely system monitors. On an ILS approach, for example, a pilot may choose to hand-fly the airplane with the aid of a flight director so that he can be prepared for system malfunctions. Since he has the plane under manual control, he is also spared the sometimes disorienting transition from a coupled autopilot approach to a manual landing.

To enjoy the advantages of a flight director, a pilot must learn how to use it properly. Flying the command bars is duck soup and presents no more challenge than does establishing an attitude on the gyro horizon. But programming the flight-director controller requires knowledge of what all the whistles and bells do. Some of the controllers have many functions, such as pitch hold, pitch sync, altitude hold, and go-around, and they can be confusing if the pilot doesn't check the system thoroughly. The need to program the flight director makes the pilot plan ahead, which helps him through his flight. On an approach, for example, he is forced to think out each element of the task in advance. He can preprogram his go-around heading with the heading bug and transition to the go-around pitch attitude by selecting that mode on his controller. He is thereby forced to think about this important but often overlooked part of the approach.

Flight directors are frequently thought of as devices only for medium to heavy twins and for turbine aircraft; until recently they have not been considered appropriate for less sophisticated planes. This thinking overlooks the principal function of a flight director, which is to determine for the pilot how much he must lead a situation in order to be in the right place at the right time. As planes fly faster and aircraft dynamics becomes more challenging, pilots have more and more trouble trying to grasp rapidly changing inflight situations. While attempting to synchronize all the information that is presented to him and to combine it with other data that he must mentally calculate, such as the rate of change of certain instrument readings, a pilot can become saturated with things to think about and do and may fail to develop the proper amount of lead. He may quickly fall behind the progress of the plane, overcontrol, and generally have a difficult time. Whether a pilot will get into such trouble depends upon many things, but foremost of these are his skill and adaptive capabilities in relation to the task before him and to the aircraft he is flying. For the businessman who may be subconsciously concerned about office matters and who does not maintain the proficiency of a corporate pilot, managing a Seneca IFR may be just as challenging as flying a Sabreliner is to a pro. A flight director does not offer much additional capability for en-route tasks, particularly since most systems are sold as an addition to an autopilot, but if the task is difficult, such as an ILS approach in turbulent conditions, or if things become confusing, a flight director can be worth a great deal, regardless of the type of plane that you are flying or the state of your proficiency.

Flight research conducted by NASA in a popular light twin has shown that a flight director can make a significant improvement in the ability of an experienced pilot to handle terminal-area and approach tasks, particularly during turbulence. They found that the command features relieve the pilot of the task of considering inputs from several instruments and leave him free to concentrate on other flight and management tasks. It seems clear that a flight director can be very useful to the man who cannot maintain the proficiency of a full-time professional pilot, and it can be a welcome addition to a single or light twin. And flight directors are easy to fly. Easiness, indeed, is their reason for being.

III. THE PILOT

Pilot technique is the concern of more than half this book; it is the crux of flying. The engineers and the test pilots see to the equipment for us; but we have to see to ourselves. The experienced pilot forgets, as he routinely executes complex maneuvers on dauntingly difficult flights, that he is unconsciously applying years of learning to a series of decisions, each of which the beginner must wrestle with for the first time.

Each author who has contributed to this book had to analyze his own actions and habits and seek the wisdom or the error that underlay them. Surprising disagreements arose; they were usually worked out, but the astute reader will detect here and there the strains of different and, on occasion, even mildly contradictory approaches to the problems of flight. It is often impossible to clearly separate a right way from a wrong way; both are part of the refinement and honing of instincts that together eventually make a pilot or a driver or an artist. Some of what is described here is protocol (such as the elaborate machinery of airspace and of IFR flying); some of it is what eventually boils down to a feeling in the seat of your pants (such as how to handle a tough crosswind landing); some of it may with luck forever remain in the realm of the theoretical for most of you (like how to make a forced landing). For the complete airman, however, none of it can be omitted.

21.

WEIGHT AND BALANCE

Gross weight is a legal expression - not an insult — and as such it tries to make precise what is really vague. Pilots are instinctively skeptical of gross-weight limitations because they sense — quite correctly — that adding 1 pound or 10 pounds or even 100 pounds to the load of a 3,000-pound airplane that is already "at gross" will not suddenly make it unflyable. Since pilots are skeptical, they may be inclined to ignore gross weight altogether. Some of us go further and ignore gross weight's evil sister, CG position. Some of us end up in fens at the ends of runways.

Gross weight is a value arbitrarily chosen by engineers to work with in designing an airplane's structure; the FAA requires certain levels of structural strength with reference to a certain load. The load is chosen by the manufacturer; the strength levels are set by the FAA; and the manufacturer tries to build a structure that will meet the FAA's load requirements and any additional criteria that experience might indicate are important. The structure should not exceed FAA requirements, since excess strength invariably represents excess weight. Furthermore, an airplane is only as strong as its weakest part, and an 11-G wing is not much use if it is combined with a 6-G tail. FAA load-factor requirements are familiar enough: normal category, 3.8 Gs; utility category, 4.4 Gs; and acrobatic category, 6 Gs. (Negative factors are generally about half of positive ones.) If your airplane claims to be a utility-category aircraft, for instance, you have some kind of assurance that its structure will withstand an acceleration of 4.4 Gs at gross weight without permanent deformation. Beyond 4.4 Gs you have no guarantee that something will not bend to the point at which it will not straighten out again when the load is removed. You still have an assurance, however, that up to 6.6 Gs — 50% above the "limit-load factor" — no part of the airplane will fall catastrophically (break, in plain English) at gross weight.

For all their to-one-decimal-place precision, FAA limit-load factors do not represent any kind of absolute facts about flight. Other countries and other administrations have used other values. For instance, England settles for 4.4 Gs for acrobatic-category aircraft, and the original value for normal category in this country was 4 Gs. The 3.8-G value is partially based on an assumed 30'-per-second "sharp-edged gust" — a model gust that is assumed to occur instantaneously. Studies of gust loads in actual aircraft in flight indicate that such gusts occur sufficiently infrequently to justify the assumption that they will not occur at all in the service life of a given airplane. If you load an airplane over gross, you nominally sacrifice the 3.8 G safety factor that is supposed to give you a certain degree of insurance against disintegrating in turbulence, but all that you really sacrifice is a certain degree of probability that you will not encounter turbulence severe enough to overstress your airplane. You have traded a 99% probability, let's say, for a 98% or 97% probability. (In fact, raising wing loading reduces gust response, and in a given gust, the least heavily wing-loaded airplane will experience the highest G loads; raising wing loading above the value at gross weight increases the actual force exerted by the gust on the wing, however, even though the G forces are reduced. In other words, the reduction in gust response does not keep pace with the increase in load on the wing. If this were not the case, a sufficiently heavy airplane would not have to have any strength at all.)

Aside from the fact that FAA load factors are somewhat arbitrarily assigned, actual airplanes are almost invariably overly strong because of a variety of engineering fudge factors and design imponderables. Some consultants make it their business to run more exhaustive analyses of the structures of existing aircraft than the factory originally ran in order to justify a gross-weight increase. The purely structural consequences of loading over gross are thus not especially significant.

The effects of overload on performance are another matter. Almost every aspect of performance − rate of descent being a salient exception − suffers from excess weight. Cruising speed, for instance, is usually reduced by about one mile per hour per 100 pounds added weight. Climb suffers more dramatically. The rate at which an airplane can climb is determined by the amount of power that it has left over from the job of merely keeping itself aloft and moving forward at a certain speed. Excess weight calls for more power merely to maintain altitude; this is trivial, however, compared to the burden that it places on the residual, or climbing, power.

One horsepower is defined as 550 foot-pounds per second, meaning that one horsepower is the power required to lift 550 pounds 1 foot in 1 second at a steady rate. Suppose your airplane weights 3,300 pounds: to lift it 1 foot in 1 second would require 6 hp. To lift it 100 feet per minute, or 1.66 feet per second, would require 10 hp. To lift it 1,500 feet per minute − the kind of sea-level climb that would be expected from a 3,300-pound airplane − would require 150 hp plus 40 hp in various losses. Your 3,300-pound airplane probably has a 285-hp engine; you can therefore conclude that it uses 95 hp just to remain airborne at climbing speed − say, 100 knots. Let's say that we have loaded the airplane to 10% over gross, for some reason; and for argument's sake, let's say that it will now climb at only 1,300 fpm. Suppose that you are starting out on a hot day at an altitude of 3,000 feet. The density altitude is 4,000 feet, and your engine is only putting out 85% of maximum power. You still need 100 hp just to stay up, and you have only 242 hp altogether. The remaining 142 hp, less losses, will let you climb at only under 1,000 fpm − a far cry from the 1,500 fpm that you expect from your airplane. If you went to 20% over gross in such a situation (it can be done), you might not be able to do much better than 600 fpm.

Even climb is not as severely affected by overload, however, as is takeoff distance. The added weight hits you two ways: it increases the time and therefore the distance that the airplane takes to reach flying speed, and it increases the speed required to become airborne. While lift is proportional to the square of speed and therefore requires only 10% more speed to lift 20% more weight, it takes quite a bit more time to accelerate from 60 mph (your normal pattern speed, say) to 66 mph (your assumed rotation speed with a 20% overload) than from 0 to 6 mph. During all that time you're sailing along the runway at 60-plus miles per hour, thus you take longer to reach flying speed, you have to reach a higher speed to fly, and you spend more time moving down the runway at high speed. The combination can be disastrous, especially if you throw in a high-altitude or high-temperature situation. Odd as it may at first seem, overweight takeoffs frequently occur precisely at high altitudes and in hot weather, perhaps because they commonly occur on vacations when the plane is loaded with family, luggage, and newly acquired knickknacks; the summer weather is warm at your mountain retreat, and you have to be back to work tomorrow and can't fool around with shuttling partial loads back and forth. Pilots sometimes feel that everything that they have to carry with them is important enough to justify the risk involved in flying overloaded, and many attach too much importance to the presumed prestige of their pilot's license in the eyes of others to say "no go" when everybody is standing around the plane with golf bags, blondes, and duty-free liquor.

If you think about the figures given above for a 3,300-pound single-engine plane with an overload, you'll quickly see that running over gross can be three or four times as serious in a twin, because it can wipe out single-engine capability entirely. Sticking to the same figures for convenience but assuming that the airplane has two engines of 142.5 hp each and that you took off with the 20% overload at the 4,000-foot density altitude and lost one engine over the end of the runway, you find that, since you have 121 hp left and need 100 hp or more to remain airborne, your residual rate of climb is 140 fpm. If the conditions seem improbably exigent to you, let's go back to sea level. Assume an average twin weighing 4,500 pounds, having 520 hp, and capable of climbing 400 fpm on one engine. It is using about 55 hp for climbing and 15 hp or so is lost, so you can assume that the rest of the remaining 260 hp (after one engine went out) is required to stay airborne. On a warm day with a 20% overload it will hardly do better than 200 fpm at sea level, and its service ceiling will be around 3,000 feet. Not so hot.

Excess weight does not seriously affect minimum single-engine control speed, though what effect it may have is for the worse. It does affect controllability in general, however, because the heavier an airplane is, the more prone it is to get behind the power curve − that is, to get into a situation in which acceleration is not possible without loss of altitude. Running over gross is not necessarily hazardous in itself − Max Conrad used to take off across the Atlantic at almost 1,300 pounds over gross in a Comanche 180 − but it brings on combinations of other problems, such as reduction of range, difficult takeoff and climb, and poor low-speed handling, which make it advisable to ponder matters with extreme caution before deciding that some circumstance or other justifies exceeding gross weight by more than 10%.

The most serious consideration in loading an airplane, discounting the poundage involved, is where you put the pounds: location can actually be much more serious than mere weight, because, while an overload makes a plane fly sluggishly, an out-of-limits CG may make it altogether unflyable. Balance, or center-of-gravity position, is only indirectly related to gross weight. Most airplanes are laid out so that increasing loads move the center of gravity backward, since the rear seats and the baggage compartment are generally behind the center of gravity. The closer you come to gross weight, the more likely you are to run outside CG limits.

The idea of CG limits arises from fundamental aerodynamic properties of the airplane. The combination of wing and tall is so designed in terms of area, position, and incidence that, as long as the center of gravity of the airplane is forward of a point called the *neutral point*, the airplane is longitudinally stable. In practice this means that the airplane has a tendency to return to straight-and-level flight if something — a gust or a control input by the pilot — causes it to deviate from straight and level. Without longitudinal stability no airplane would be capable of hands-off flight: any slight nose-down deviation would generate a dive, and any pitch-up, however slight, would lead to a stall. An airplane with neutral longitudinal stability would be flyable and one with slight longitudinal instability might be controllable by a good pilot, but anything but positive longitudinal stability would be uncomfortable at best and dangerous in any case.

The neutral point on most airplanes is located around 35% to 40% of the wing chord aft of the leading edge. CG limits of 15% to 30% of chord are commonly permissible, meaning that the CG can be anywhere between 15% and 30% of the way from the leading edge to the trailing edge of the wing, and the plane will retain satisfactory flying characteristics. The forward CG limitation is determined not by stability but by controllability: stability increases steadily as CG moves forward, but the ability of the horizontal tail to generate enough force to flare the plane for landing decreases as the CG moves forward. Cessna's Cardinal, a comparatively nose-heavy airplane when lightly loaded, has a powerful all-flying horizontal tail to handle the job of flaring to land. On a stabilator-equipped aircraft flaring power may be increased by trimming the nose down: the increased stick force necessary to flare reflects increased effectiveness in the stabilator. In any case, however, few airplanes permit a CG position much ahead of 10% of chord.

Fuel tanks are usually located ahead of the center of gravity for reasons involving the flexural properties of wings and because in low-wing retractables the wheel wells are located aft of the main spar. Fuel burnoff therefore causes the center of gravity to move backward, and it is possible in some airplanes to take off with an acceptable CG position and then to have the CG migrate out of the safe envelope during flight. This could be an extremely dangerous condition, especially if it were combined with an overgross loading with its implied rearward concentration of weight. It is for this reason that handbooks and instructors always have you calculate the CG for both takeoff and landing on any trip on which the load may be doubtful. Like overgross loadings, out-of-limits CG positions are something about which pilots tend to be casual, because they come up so rarely (how many times have you heard a pilot boast that he has never done a weight-and-balance computation since he got his ticket?), and because pilots sense that running an inch-pound or two out of the envelope is not likely to suddenly sink us. Caution, however, never killed anybody. And, after all, just how much lousy performance and instability are you willing to cope with in order to get your friends' adobe amphora back from Mazatlan?

22.

GETTING OFF THE GROUND

Takeoff distance is determined by a number of factors − not so many as to put the problem beyond the reach of simple computation but too many to juggle in your head while you're mentally sizing up the runway, altimeter, and thermometer. For years books on how to fly provided us with something called the *Koch chart* for determining takeoff distance on the basis of density altitude and the airplane's rated performance, but it got used about as often as weight-and-balance schedules. The FAA then offered a substitute for the Koch chart: the Denalt performance computer. It comes in two varieties − one for use with constant-speed propeller and the other with fixed-pitch − and it is used to find expected takeoff distance and rate of climb, given certain conditions of pressure, altitude, and temperature. It is an improvement in packaging over the old Koch chart but is surprisingly poorly produced, considering that the whole resources of the United States Government went into it. Not that it isn't well printed or is made of an ecologically unsatisfactory type of plastic, but it contains, of all things, mistakes. There are several, sprinkled throughout the computers, and they betray themselves to a careful eye as nonlinearities in the answers that the computers give. For instance, on the fixed-pitch model for an air temperature of 60° F the factors by which you should multiply rated takeoff distance to get expected performance at 2,000, 4,000, 6,000 and 8,000 feet are, respectively, 1.4, 1.9, 2.1, and 3.0. The 2.1 is obviously wrong: it should be 2.4 or 2.5. There are a number of such mistakes in both computers.

In spite of the errors, however, the Denalt computers are handy to have around, since they provide you with a shortcut to the answer to a performance problem that, if it is a problem, may be absolutely critical. As with most handy gadgets, however, the essence of their proper use is to use them: all too often pilots invest in some nifty little toy and then find that fishing it out of a flight bag at the proper moment is more trouble than doing without it. Perhaps the best way to persuade you of the importance of the Denalt computer is to explain how altitude and temperature affect takeoff and climb − and leave you to imagine how the trees at the end of the runway affect airplanes.

The sine qua non of flight is air. There is but one kind of air, for all practical purposes: it is more or less chemically uniform throughout the lower atmosphere, but its density varies from place to place on and above the surface. Density means the mass − or weight, if you prefer − of air in, say, a cubic foot of it. To put it another way, the density is the number of molecules found in a given volume. Density is a function of pressure and temperature. It may be hard to see intuitively how density, pressure, and temperature interact in the atmosphere, but it is sufficient to know that increasing pressure increases density (by squeezing the molecules closer together more of them fit into a cubic foot). Decreasing temperature increases density as well (by slowing the movements of the molecules it is possible to fit more of them into pressure, just as one could probably fit more dead or moribund bees into a milk bottle than live and angry ones).

Pilots find themselves faced with a verbal complication in the term *density altitude*, but density altitude is simply a convenient way of expressing temperature and pressure together by reference to a list of densities called the *standard atmosphere*. The standard atmosphere is an idealized profile of the earth's atmosphere, starting with the familiar sea-level "standard day", with a temperature of 59° F and a barometric pressure of 29.92" of mercury, and going on up to heights where there is one molecule to a cupful of space and temperature isn't temperature any more. In effect, the standard atmosphere is simply a list of all possible air densities that may be found on earth, and it makes it possible to express density in terms of an altitude rather than in terms of pounds per square foot. Density altitude, therefore, is the altitude in the standard atmosphere at which the air density is the same as it is at the place and time for which the density altitude is specified. Note that if the pressure is standard but the temperature is above standard, the density altitude will be above standard; if the temperature is standard but the pressure is above standard, the density altitude will be below standard. Most pocket flight computers can also be used to find density altitude from temperature and barometric pressure.

Insofar as pressure may be measured without reference to temperature, there is a separate quantity called *pressure altitude*. It is similar to density altitude in that it represents a reading of current pressure in terms of the pressure profile of the standard atmosphere. If you are at sea level and the barometric pressure is a sunny 30.91, the pressure altitude is below sea level. In order to read a pressure altitude of sea level, the barometric pressure would have to be 29.92. While density altitude is found on the computer, pressure altitude may be found by setting the barometric-pressure adjustment on the sensitive altimeter to 29.92; the altitude indicated on the instrument is now the true altitude, which is shown if the number in the window is the true local barometric pressure, as supplied by, for instance, a flight service station.

This raises a question as to the basic conditions of the takeoff-performance problem. Takeoff performance depends upon thrust, weight, rolling friction, wind, and runway gradient. Assume for the moment that there is no wind, that the runway is level, that you are at gross weight, and that the runway has a blacktop surface. In order to take off with a given weight, the airplane must be moving at a certain indicated airspeed — say, 1.1 times its flaps-up stalling speed. It must be an *indicated* airspeed, because the wing "feels" indicated speed, and true airspeed is figured out later on. If the density altitude is sea level, the airplane is guaranteed by the manufacturer to be able to clear a 50-foot obstacle in 1,500 feet from the start of the takeoff roll. It is up to you, however, to figure out how it will behave if the density altitude is not sea level — which happens practically all the time. Most of the time it hardly matters: you have plenty of runway, and altitudes and temperatures are moderate. From time to time, however, almost everyone encounters a borderline situation — "hot and high" and a short runway. "Hot and high" means that the pressure altitude and temperature are both above standard and that the density altitude is therefore considerably above standard. There are fewer gas molecules per cubic foot than you have come to expect.

The lack of air affects the airplane in three ways. First, it prevents the engine from developing full power. Since the engine derives its energy from burning gasoline and air in a fixed proportion and its ability to suck in air is limited by ambient pressure, engine speed, and intake configuration, the less dense the ambient air, the less fuel the engine will be able to burn and the less energy, or power, it will put out. At a density attitude of 7,000 feet, for instance, a normally aspirated engine will typically give only 75% of its sea-level power.

Next, a lack of air affects the propeller. Since engine speed is limited by the redline, the true airspeed of the propeller blades cannot exceed a certain preordained takeoff value. In air of less than sea-level density, the propeller is thus traveling at, so to speak, a reduced indicated airspeed: that is, if you had a pitot tube at the propeller tip, it would read lower than at sea level. Since thrust is proportional to the square of speed, the loss of effective speed − or of dynamic pressure on the blades, to be exact − may sharply reduce the *usability* of the already curtailed engine power. It is in fact because of this combination of engine-power loss with altitude and blade-speed limits that helicopter performance is so sensitive to density altitude.

Finally, high density altitude affects the airplane's wing just as it affects the propeller, by reducing its effective forward speed. If the airspeed indicator says 70 mph at a density altitude of sea level, you are going 70 mph (ignoring pitot-system error). At a density altitude of 5,000 feet, however, if the airspeed indicator says 70 mph, you are actually going 75 mph − but only feeling 70 mph's worth of air going by the airplane. If you were taking off and needed to indicate 70 mph before rotating, you would then have to accelerate the airplane to a groundspeed of 75 mph.

All these effects compound one another. To take off at high density altitude, you have to get the airplane's mass up to a higher speed than usual and you have less power than usual with which to do it. The result is that the takeoff run takes more time, and, since a lot of time is spent rolling along at a good clip, it may take quite a bit more distance. Add to this the effects of overweight (extra weight means a higher liftoff speed and slower acceleration), runway friction (an unimproved strip with tall grass or soft ground may have a rolling resistance 10 to 15 times as great as that of a blacktop runway), and possibly a runway gradient (mountain runways often slant, and sometimes the terrain forces you to take off in one direction regardless of the slant or of the wind), and you have the makings of a disaster.

Going back to the example airplane with its 1,500-foot advertised takeoff performance: if it has fixed-pitch prop, it will require about 2,400 feet to clear a 50-foot obstacle at an altitude of 2,000 feet and an air temperature of 80 F. At 4,000 feet and the same temperature it will need 3,000 feet. In a really extreme case − say 6,000 feet high with an OAT of 100 C, (a situation that one might easily encounter in Mexico) − the distance goes up to 4,500 feet. With a constant-speed propeller the situation is not quite so grave: in the three examples just cited, the distances are 2,100 feet, 2,700 feet and 3,800 feet.

Taking off into the wind will reduce the takeoff distance but will not increase

116

the rate of climb. (It will, however, increase the angle of climb.) Taking off uphill may considerably increase the takeoff distance. It must also be borne in mind that factory takeoff distances represent the best that the airplane can do in the hands of a pilot of "average competence," a proviso that the manufacturers are not likely to interpret to their disadvantage. They also represent the best that a clean, perfectly rigged, new airplane with a new engine can do. The power output of a reciprocating engine drops with age in almost every case. Finally, reduced rate of climb may kill you even if you succeed in taking off. A sea-level eyeball estimate of the possibility of clearing this or that obstacle will be way off at high density altitudes.

A last complication: since you are using a higher-than-usual groundspeed to take off at high density altitudes, you will take more distance to stop if you decide to abort the takeoff. Most of this applies principally to single-engine airplanes, since some twins face severe loss of single-engine capability even at fairly modest altitudes, but the problem of aborting the takeoff is as critical to twins as to singles. At high density altitudes, it is well to keep in mind that the loss of half the power may put you in exactly the same fix as the loss of all of it. You should get the appropriate Denalt computer for your airplane (at a pilot's shop or by sending 50 cents to the Superintendent of Documents, U.S. Government Printing Office, Washington, D.C. 20402) and keep it in your flight bag. If in doubt, use it − but check the numbers for logical consistency, since there are several errors on the computers. If, after checking out the situation on the computer, you're still in doubt, don't go. Remember, luck isn't *always* with you.

23.

SOME FINE POINTS OF TAKEOFF TECHNIQUE

If you had a dime for every word that has been written about landing airplanes, you could probably buy yourself a shiny new airplane. If you had a dime for every word written about taking off in airplanes, you might just be able to buy a radio or two. You often hear pilots brag about that perfect landing, but you seldom hear a pilot say he lifted off and climbed with the grace of an eagle. Possibly the feeling that the takeoff is something that "just happens" has led to a lack of interest in the subject. Takeoff complacency may even have contributed to the fact that there are nearly as many serious general aviation accidents in the takeoff and initial-climb phases of flight as during approaches and landings. Furthermore, while mechanical problems should be the only reason for marred takeoffs, only about 9% of takeoff mishaps occur because something on the airplane actually broke. The National Transportation Safety Board, in its computerized accident summaries, tells us a lot about takeoffs. For instance, the takeoff run is a simple maneuver, and the fact that only 20% of the takeoff accidents occur during the run indicates that pilots generally have mastered it reasonably well. Most of the failures involve loss of directional control, usually in a strong crosswind, though some pilots manage to lose control even in calm conditions.

Perhaps the most important thing about takeoff-roll technique is to relax on the rudder pedals. As the airplane accelerates, some pilots become tense and press both pedals at once, inducing a sort of self-generated friction into the steering system. This can lead to abrupt overcorrections while tracking down the runway and to loss of control. Such accidents appear to occur most frequently in winter and early spring when crosswinds are strongest. Snow piled beside the runway both induces and softens the stop of the wayward plane. The commonest way to make a good crosswind takeoff is to line up on the white stripe, lay the lash to it, steer with the pedals, and, by applying aileron into the wind, keep the wings level and prevent any sideload from developing. The airplane should be allowed to accelerate to beyond the normal liftoff speed to ensure a clean liftoff with no settling back onto the runway.

One mistake that can lead to directional-control difficulties during a crosswind takeoff is to keep all the wheels on the runway past normal rotation speed by pushing the control wheel forward. Holding the wheel forward to keep the airplane on the ground can directly affect the weight distribution on the wheels. The stabilator or horizontal stabilizer and elevator actually tend to lift the tail off the ground, reducing the weight on the main wheels and concentrating it

118

on the nosewheel. This can lead to the same sort of wheelbarrowing that occurs if a pilot attempts to push the nosewheel onto the runway and hold it there with the elevator or stabilator after touchdown. There's just no way to control an airplane whose weight is concentrated on the front wheel. The moral of all this is that the pilot should understand his airplane's landing-gear geometry and its relationship to weight distribution on the wheels.

The taildragger pilot has the easier job. He can raise the tail of his airplane early in the run to concentrate the weight on the mains, run to a good liftoff speed, and fly away. The pilot of most tricycles can't do anything to put more weight on the mains than he has at the start − he can only work at not putting less on them. The tricycle that rests at a neutral or slightly negative attitude is best on a crosswind takeoff − you can run it up to a good speed without it becoming light on its feet. A tricycle that rests slightly nose-high is most difficult to handle in a strong crosswind: it becomes lighter on its feet the faster it goes, and trying to keep it on the ground only makes the problem more acute.

Another method of taking off in a crosswind, not commonly used but very effective if the runway is reasonably wide, utilizes a curved run to take advantage of centrifugal force in offsetting the effect of the crosswind. The run is started on the downwind side of the runway and is pointed 15 to 20 degrees off the runway heading and into the wind. As the airplane accelerates, a very gradual turn toward the runway heading is begun so that the run describes an arc from the starting point to the upwind edge of the runway. If liftoff speed has not been attained at that point, the gentle curve is continued. If you want to quit when the airplane reaches the runway heading and revert to the normal method of crosswind takeoff, that is fine. The curved takeoff roll is something that should first be tried in the company of an instructor who is familiar with it. It is a good procedure if done properly, but it does demand an understanding of the airplane's performance. A pilot who doesn't know how far his plane is going to run before it will fly might make too sharp a curve and wind up in the ditch.

Lifting off at the proper time is an important part of the takeoff. In fact, the initial climb is where most takeoff accidents occur. There is a best speed for every takeoff, with weight being the prime variable in the case of single-engine airplanes. Minimum single-engine control speed often determines the takeoff speed of a twin, regardless of the operation weight. A good guide for a single is to use a takeoff speed 10% above the power-off stall speed (1.1 Vs) for the operating weight. In case of power loss after liftoff this ensures some margin of safety. Some overzealous pilots will take off at lower speeds to demonstrate their prowess, but it is dangerous to do so.

Stall speeds for gross-weight operation are listed in the aircraft owner's manuals. The Cessna 172, for example, stalls at 57 mph, so for takeoff add 10% to that and begin a liftoff at 63 mph. If you were operating with two people and half gas aboard or at 450 pounds (approximately 20%) below gross on an average 172, the stall speed would be approximately 10% lower (half the percentage of the weight below gross). With a 51-mph stall speed, the liftoff could begin at 56 mph with good margin. If the field is short or soft or if the grass is tall, it could be important to calculate this and take advantage of an earlier liftoff. Nobody overloads airplanes, of course, for it is against the law, but if someone did, he would hopefully remember that overloading raises the stalling speed above the published gross-weight figure and that the liftoff speed should be increased. The increase should be equal to half the percentage over gross.

Trying to take off too soon is fraught with peril, because pilots often try to do it when the airport is short — the looming trees make one's instincts suggest that is is time to go up — and premature liftoff actually prolongs the takeoff roll and cuts initial climb. On a good surface, the airplane will accelerate better if it is running level than if it is running along with the nose up, because there is more aerodynamic drag in the latter condition. If the pilot does succeed in wishing the airplane off the ground at a low speed, the aircraft will be flying at a very high angle of attack, and drag will be maximized. There will be no way to make the airplane climb without first accelerating, which consumes distance. The attitude of the airplane can tell you a lot as it lifts off the runway. In a proper takeoff, the attitude at which the airplane first flies should be approximately the same as that at which it climbs. If the nose must be lowered after liftoff to assume the climb attitude, liftoff was sooner than it should have been.

Accelerating to a higher speed than necessary is wasteful but not as hazardous as a premature liftoff. Most pilots accelerate to higher-than-necessary speeds only if there is plenty of runway, so the prime loss would be in simple efficiency plus a little wear and tear on the tires. If the takeoff surface is soft, there is an immediate necessity both to lighten the load on the wheels during the run and to fly the airplane. The drag increase is still there, but it probably doesn't equal the drag caused by rolling the relatively small tires through mud, sand, or tall grass. The smart pilot remembers that there is no such thing as a combination soft/short-field takeoff procedure. You can do one or the other but not both. Unfortunately, many short fields are also soft.

The first 10 to 20 degrees of flap deflection usually increase lift more than drag, lowering the stalling speed without any great drag penalty. To leave a small airport, it is usually best to use the maximum-lift flap setting. The airplane owner's manual will be specific on this, and its instructions should be followed. A few airplanes do best with no flaps extended: if you are flying one of these, don't be tempted to try a little flap just because it works on other airplanes.

After lifting off the runway into the initial climb, find the best-angle-of-climb speed, Vx in engineering terms, which is the velocity at which the airplane will deliver the most "up" for the forward distance traveled. This speed should be given in the airplane manual. If it is not clearly identified, consult the takeoff-distance chart: the speed used for climb is the best-angle-of-climb speed. The published climb speeds are for gross weight unless otherwise specified and can be adjusted for climb at light weights for maximum efficiency. The rule of thumb here is the same as for reduction of stalling speed at reduced weight: cut the best-angle-of-climb speed by half the percentage under gross weight. If the speed is 90 mph and the operating weight is 10% under gross, shave five percent off 90 and climb at about 85 for best angle.

The best-angle-of-climb speed is critical if obstacle clearance is involved. If the airplane will make it over the trees, it will do so at that speed; if it can't make it at that speed, it won't make it at any other speed. Going slower is especially hazardous, since the best-angle speed is not too much above stall on some airplanes.

A pilot must be aware of the attitude that will yield best angle of climb. The airspeed indication tells you whether the attitude is correct and what to do about it if it is not correct. A special place to beware of poor attitude information is in taking off toward rising terrain or a mountain. The horizon will be below the highest terrain or at the base of the mountain, which can prove very confusing to flatlanders. On such a takeoff the artificial horizon would provide the most reliable attitude indication, and the airspeed indicator should be checked frequently to make sure that the correct attitude is being flown.

The next speed to consider is best rate of climb, or Vy. Once the obstacles are cleared, this speed should be used while you are first gaining altitude, for it will yield the highest possible rate of climb. Best-rate speed may be reduced at lower weights, using the same rule of thumb as before.

The actual takeoff in a twin is somewhat different from that in a single, although the basic principles remain the same. The only additional thing to consider about the twin is the minimum single-engine control speed or Vmc. This is the lowest speed at which the airplane can be controlled with one engine wide open and the other windmilling. Not easily controlled, necessarily, and there's no guarantee that the airplane will climb or even fly level at that speed. This speed should be used like stall speed in a single − as a reference only. It is best to plan the liftoff at a speed 10% above Vmc or 10% above the stall if that value should happen to be higher.

The speed that really counts in a twin is the best engine-out angle-of-climb speed. Until the airplane reaches that speed, there is little chance of staying in business in case of an engine failure. This is the speed around which a pilot should base all his thinking during takeoff in a twin. Below that speed the takeoff has little or no chance of success in case of engine failure, and the takeoff should be aborted regardless of what is ahead. Above that speed chances are better. The speed should be used as in a single: fly it until the obstacles are cleared and then use the best-rate-of-climb speed.

There has always been much discussion on how best to fly the initial climb in a twin if both engines are running. Some like to zoom upward at the best-rate-of-climb speed, putting altitude in the bank as quickly as possible. Others prefer using a little extra speed so that any speed deterioration that might accompany an engine failure wouldn't put the airplane in a position of being below best-climb speed and of having to accelerate to the best speed for single-engine flight after an engine failure. The altitude theory is better and more efficient than the airspeed theory, given flawless pilot technique, but when average pilots are considered, it's probably six one way and half a dozen the other. The twin pilot really negates his investment in two engines if he doesn't carefully consider each takeoff and weigh all the factors. Unfortunately, many twin takeoffs are made in situations where the airplane would not have a chance of making it around the airport if an engine failed in the initial-climb phase of flight.

There is something that will cure a high percentage of takeoff woes − single or twin. It can be done easily, will lead to perspiration-free and safe takeoffs, and, in the case of the twin, will give the maximum advantage to the investment in two engines. The magic thing is something that you can learn from the airlines. You don't often hear of them having takeoff problems, and the reason for this is that they meticulously plan each and every ascent. Weight, wind, and temperature are applied to the airplane's performance capability, and, if it won't go with specified and generous margins, they don't go. The general aviation pilot needs only to plan like the airline pilot to imitate the latter's good takeoff record.

The first part of the plan recognizes that warm air is less dense than cool air.

This basic phenomenon plays hob with aircraft performance, but you can plan around it through the use of density altitude − altitude compensated for temperature. Your airplane's capabilities are expressed in relation to density altitude, not to the altitude shown on the altimeter, and the pilot who doesn't make allowances for above-standard temperatures is courting disaster. Many pilots like to fly in high country in the summertime. The density altitude at Denver when the temperature is 85 F is 8,000 feet. What is the maximum full-power lean-forward-in-the-seat rate of climb available to your airplane at that density altitude? A 300-hp Cherokee Six will struggle upward at only 575 fpm under these conditions, little more than half what it does at sea level at any standard temperature. Move over to Aspen, cool things off to 70 F, and try again. The density altitude is 10,000 feet, and the maximum available rate of climb is 475 fpm. Turbulence would probably cut those figures by 50 fpm, and a downdraft could easily put them in the minus column. Those figures should also be shaded somewhat because an old engine may not be as peppy as it was when Lycoming first shipped it to Piper. The actual takeoff-distance figures in the owner's manual only go up to 7,000 feet density altitude (at which point they show almost 3,500 feet over a 50-foot obstacle), so you'd be on your own in calculating which distance would be required at higher density altitude. The answer would be a lot, and the message would be to wait for cooler weather or to lighten the ship as much as possible before starting out. By cutting the load you can figure on improving performance by a greater percentage than the reduction below gross weight. To be on the safe side, just count on cutting takeoff distance and increasing rate of climb by the same percentage. What should the pilot flying a twin in high-density-altitude conditions do? Aspen on a warm day is far above single engine ceiling of most twins, as is Denver. The weight situation of many airplanes is such that operation at reduced weight means either little gas or few people. The answer is simple: either don't go or operate the twin as if it were a single-engine airplane. If one quits, shut the other one down and land. Even lower-elevation airports can pose density-altitude problems on hot days. Teterboro, New Jersey is about as low as you can get in relation to sea level, yet a muggy summer day can run it up to over 2,000 feet. High humidity also cuts engine power output. The first part of the plan is thus to include the temperature in your calculations, to be certain that adequate runway is available for the takeoff, and to know that your rate of climb will be sufficient to clear obstructions.

The second part of the takeoff plan comes in the preflight inspection and run-up. A detailed checklist should be used for every takeoff. Make sure that there's no water in the fuel; select a tank with fuel in it; check the mags; run the fuel pump if appropriate. Make sure that the controls are free and correct. It has happened that during major repair work control cables have been incorrectly hooked up.

The third part is to plan how the takeoff will be conducted. Liftoff speed, for example, might be 65 mph and best-angle-of-climb speed 87: simply make sure that you break ground and climb at those speeds.

The last part of the takeoff plan comes into play as you start down the runway. It is the abort plan. Any abnormal noise or indication should be sufficient to chop the power and stop before it is too late. Noise can tell a lot. An improperly latched door will often begin saying something early in the takeoff run, for example, and it's best to step on the brakes and shut the door. If you happen to be flying a retractable, there is an open-door trap that you must watch out for. More than one pilot has taken off, put the gear switch in the up position, had the door pop open, and decide to land on the remaining runway to close the door — without lowering the gear.

Be alert for sluggish acceleration early in the run. If it is a short airport and if your calculations of the required distance to fly are based on a hard surface, the first sag in acceleration due to surface softness should give the word to abort. So should any other feeling of slow acceleration. Some people actually time takeoffs to satisfy themselves that they are getting underway properly. To figure takeoff-roll times, correct the indicated airspeed at liftoff to true airspeed; then cut that in half and convert it to feet per second for average groundspeed during the run. Divide the groundspeed in feet per second into the calculated takeoff roll in feet, and you'll end up with the number of seconds that it should take to get airborne. This might sound a bit approximate, but takeoffs are usually within a few seconds of the calculated time. Some pilots use a system of picking a point along the runway as a go/no-go guideline on takeoff. If liftoff speed hasn't been reached by that point, they abort.

The planning that leads to a good, safe takeoff every time takes but a few minutes, and the execution of the plan is not something that takes a great amount of skill. There's nothing difficult about consulting the takeoff-distance chart. The speeds to use are also in the book. And it is easy to lift off and climb at the correct speeds. Do it right and avoid the problems that arise when a pilot tries to make an airplane do something that it cannot do.

24.

RICH, LEAN

You'd never guess from the small size of the mixture control and the short shrift given it in flight training that it is one of the two or three most important engine controls in the airplane — along with the throttle, say, and the ignition switch. It doesn't look like much, but the mixture plunger is the key to major improvements in fuel economy and time between overhauls if properly used and the shortest cut to catastrophes of several sorts if you use it wrongly. The word *mixture* refers to the blend of air and vaporized gasoline that the engine burns. There is a "chemically correct" mixture — 15.2 pounds of air per pound of gasoline — at which both air and gasoline are completely consumed, leaving "pure" combustion products (carbon dioxide and water). Mixtures as rich as 9:1 and as lean as 20:1 are explosive, however, and produce, in addition to the combustion products of the ideally matched portion of the mixture, various "products of dissociation" — of which carbon monoxide is a familiar example.

A "rich" mixture is one containing an excess of gasoline. A "lean" one contains excess air. These nonideal mixtures may be selected for certain reasons — cooling, for instance, or producing maximum power. Mixtures richer than the ideal are usually used and leaner ones avoided, though Lycoming has authorized operation "on the lean side of peak." "Peak" refers to the "peak exhaust gas temperature," which is achieved with the chemically correct mixture; since the peak temperature signals the correct mixture, rich and lean mixtures are measured for convenience in terms of degrees Fahrenheit below peak on the rich or lean side. When Continental recommends cruising at "25 degrees to the rich side of peak EGT," they mean that you should maintain a mixture containing enough excess gasoline to lower the exhaust gas temperature by 25 degrees F.

In addition to the cockpit mixture control the engine is equipped with several systems to vary the mixture automatically. Carburetors and injection systems are designed to enrich the mixture at idle and at full throttle and to lean it automatically in the normal operating range. The enrichment at idle is necessary in order to counteract the bad effect on combustion of unscavenged exhaust gases from the last power stroke in the cylinder; enrichment at full throttle, on the other hand, is intended to cool the engine and to protect it from damage due to detonation. This protection is provided by an automatic enrichment valve. While the engine's automatic mixture control systems are satisfactory to control mixture throughout a range of throttle settings, they do not make adjustments for different ambient-air densities (except on the larger-geared Lycomings, such as the ones used in Queen Airs, Twin Bonanzas, and older Aero Commanders). This is where the pilot's mixture control comes in: it permits the pilot to select the desired EGT at any density altitude and also to vary the mixture for specific purposes. The full-rich mixture setting on a typical light aircraft is chosen to give a satisfactory mixture for cooling and for avoiding detonation during takeoff and climb. At full power it is necessary to maintain an EGT as much as 200 degrees below peak for safety; this is where the enrichment valve comes in. In very cold weather, however, some aircraft (Cessna 180s and 182s, for instance) run abnormally lean even with the mixture control all the way in, and it is necessary to reduce the density of the intake air by applying carburetor heat in order to ensure safe engine operation. Thus the carburetor heat control may be thought of as a kind of continuation of the mixture control.

Permissible EGTs depend on engine power settings, because engines are stressed to withstand certain combinations of temperature and cylinder pressure. The higher the "mean effective pressure" in the cylinders (that is, the higher your power setting), the lower the temperatures that the engine can handle without overstress or detonation. It has been customary in the design of light-aircraft engines to place the power limit for lean cruise operation at 75%, though it is 65% in some Lycoming supercharged engines. This means that the engine should be run at full-rich mixture above 75% and may be leaned at any time or altitude if it is delivering 75% or less. In practice throttle settings between full power and approximately 80% are not used — except insofar as altitude may turn a full-throttle setting into a power level of less than 100%. An important and little-recognized rule about climb procedures, in fact, is that throttle reduction in climb should not be attempted unless the reduction is all the way back to a setting recommended by the powerplant manufacturer for climb power (usually between 75% and 80% power). Many pilots, after using full throttle for takeoff, think that they are doing their engines a favor by throttling back slightly for climb. In fact, they are making it harder on the engine by depriving it of the benefit of the mixture-enrichment valve. By slightly reducing throttle after takeoff you lean the mixture and may push the EGT to an excessively high level. The correct procedure would be either to maintain full throttle until arriving at an altitude at which you can comfortably throttle back to 75% or to throttle back to 75% immediately after takeoff.

Though it is given to students as a rule of thumb that they should not bother with leaning the mixture below 5,000 feet, the truth is that the mixture may be leaned out at any altitude, provided that the power setting is 75% or below. "Conservative" use of mixture — such as refusal to lean below 5,000 feet — is in fact unconservative of both fuel and spark plugs (the latter are prone to foul in rich mixtures).

In leaning by ear, the best one can do is to lean the engine until it shows a loss of rpm or begins to run a little rough and then to enrich the mixture until smooth operation is restored. It is not necessary to restore peak rpm (in fixed-pitch aircraft; with a constant-speed prop rpm does not vary significantly during leaning); in fact, peak rpm occurs at "best power" mixture, which is about 100 on the rich side of peak EGT. The reason why a mixture richer than the chemically correct one produces — more power than the chemically correct one does is that the products of dissociation act to increase mean effective pressure over that obtained in "pure" combustion at 15.2:1. The cost is high, however: for an increase of 1% or 2% in speed you sacrifice about 14% of your range. A mixture is set for only one combination of manifold pressure, rpm, and density altitude. Changing any of these factors requires a change in the mixture − though the change may be too small to be worth bothering with. Don't be hesitant about playing with the mixture, however: even if you lean excessively, the engine will not stop. In fact, it will windmill even with the mixture pulled all the way out to idle cutoff.

The roughness that you feel if you get on the lean side of the proper mixture is due to inequality in the feed to the different cylinders: the leanest cylinder is obviously the first to sputter if you lean excessively, and it is the cylinder for which you should try to set the correct mixture. All the others will be to some degree richer.

The effectiveness of the mixture-control plunger is nonlinear: that is, it may not do much for 1" and, it may then do a great deal in the next 1/8". Leaning by eye is therefore not advisable. If you don't have an EGT gauge, engine sound and rpm are the only reliable clues to proper mixture. You typically take off with full power and full-rich mixture, unless you are operating at so high an altitude that leaning is necessary for smooth operation at full power. At a certain point you would throttle back to 75% of power and then lean your mixture in order to stay *on the rich side of best power* (or peak rpm) until you level out. When you level out, you select cruise power and lean to peak EGT, or to smooth operation *on the lean side of best power*, or to 25 degrees on the rich side of peak EGT, whichever you prefer.

If you use carburetor heat during the flight, you have to readjust your mixture accordingly: carburetor heat, remember, *enriches* the mixture. If you fly through a front, say, into markedly warmer or colder air, you must also readjust your mixture. Curiously enough, you must pay particular attention to mixture in flying over water, over rough terrain at night, or in any other situation in which you might be nervous and particularly attentive to engine sound: even slight engine roughness (which is natural in piston engines) tempts most pilots to enrich the mixture a little for safety's sake, which not only increases the chance of plug fouling (more roughness, more enriching) but also makes surprising inroads into range. If you have been flying at a moderately high altitude (say, 7,500 or above) at full throttle and you reduce throttle to descend, you should reset the mixture, because by reducing from full throttle you have in effect leaned the mixture by cutting out the enrichment valve. On the other hand, you can afford to run abnormally lean during descent, since there is little or no danger of overheating.

The loss of fuel economy − and therefore of range − that can result from improper leaning is amazing: for instance, an airplane may consume twice as much fuel at full-rich mixture at 7,500 feet as at full-lean and therefore fly only half as far. If you have been disappointed by the failure of your fuel consumption to tally with that given by the manufacturer, your mixture settings are most likely at fault.

Some people are leery of operation at lean mixtures because, if you run at too high a power setting (or too high a manifold pressure combined with too low an rpm) with an extremely lean mixture, you may overheat, damage your exhaust valves, and/or cause detonation in your engine. What the overcautious ignore, however, is that by running at too rich a mixture you waste a fortune in fuel, subject your spark plugs to more rapid fouling, and increase air pollution. In this last respect, one airplane is worth several cars.

Detonation is the less dangerous of two kinds of "knock" that can occur in a hot engine at high cylinder pressures. The other is preignition. In detonation, a portion of the fuel-air mixture in the cylinders spontaneously explodes ahead of the wave front of combustion previously ignited by the spark; this secondary combustion may be triggered by high engine temperatures and high pressures within the cylinder, which in turn may be produced by a lean mixture combined with high manifold pressure. The effect upon the piston is as if it were being struck by a hammer rather than by a rubber mallet, and, though detonation may continue for some time without causing failure, it wears and fatigues the piston more rapidly than would normal combustion. Preignition, on the other hand, is premature ignition of the entire fuel-air mixture early in the compression stroke by an incandescent particle or projection in the cylinder − a rough edge on an eroded piston, for instance, or a bit of carbon pried up from the piston by the shocks of detonation. If preignition occurs, combustion-chamber cooling practically ceases, and complete failure of the piston or cylinder may occur within less than a minute. While detonation may sometimes be audible in an automobile engine as a sort of pinging or rapping noise when you step on the gas or go up a steep hill, it is inaudible in an aero engine, so avoidance by precise and intelligent mixture control is especially important. Using the correct grade of gasoline is, of course, even more essential: no amount of mixture control will prevent detonation of a high-compression engine at even moderate power settings if it is fueled with 80-octane gasoline.

While mixture may be adequately controlled by ear and somewhat less adequately with a fuel-flow gauge (they are notoriously inaccurate) and while the cylinder-head temp gauge may serve in a sort of advisory capacity for mixture adjustments during climb, there is no substitute in the long run for an EGT gauge. An EGT gauge that reads the leanest cylinder or a more elaborate arrangement that can read any cylinder at choice is not expensive to start with, and it quickly pays for itself in savings in fuel, spark plugs, and perhaps even valves. It may substantially increase time between overhauls, and a sophisticated gauge-reader can detect the onset of any of a variety of ills in the engine just from the exhaust temperatures in different phases of operation and from the rate at which temperatures change with mixture adjustments. An EGT gauge also provides a handy way of monitoring the effectiveness of the mixture-enrichment valve: some carburetors, particularly certain pressure carburetors, may deliver excessively lean mixtures at full throttle − a circumstance that, without an EGT gauge, may be discovered only after a premature overhaul.

The little mixture control that many pilots treat so casually is really the only way that you have of choosing between the right fuel and the wrong fuel for each power setting and density altitude. If you take the time to set up your mixture carefully in the first place and to readjust it each time your power setting or the density altitude changes or you use your carburetor heat, you will be amply repaid in improved performance, better fuel economy, longer engine life, and, *Deo volente*, maybe even a cleaner environment.

25.

WIND, SPEED, AND RANGE

By now, everyone should know that on a round trip any wind (unless it changes direction in your favor during the trip) is an ill wind. A headwind one way doesn't cancel itself out by being a tailwind on the way back. The reason is that you are affected by the tailwind for less time than by the headwind. Take a simple example: you fly a 100-nm trip with a 20-knot headwind and return immediately with a 20-knot tailwind; your TAS is 100 knots. On the way out, your groundspeed is 80 knots; returning, it is 120. Going takes 75 minutes, and returning takes 50 — a total of 2 hour and 5 minutes, which is 5 minutes more than if there had been no wind at all. No matter what combination of trip length, wind speed, wind direction, and airplane speed you choose, the result is always the same: if there's a wind, you lose.

What about a sidewind, you wonder? A wind blowing at 90 to your course resolves itself into a small headwind; in order not to affect groundspeed at all, the wind must be blowing from slightly behind the broadside position. If you have such a wind while flying one way, however, it becomes a headwind when you turn about, so again you lose. The closest you can come to beating the game is to fly different routes coming and going in order to make the best of local variations in wind direction.

Suppose that you were to increase your TAS while flying into the headwind in order to improve your groundspeed and reduce your time en route and then to throttle back on the return leg. Could you cruise at 120 knots going and 80 coming back for a uniform groundspeed of 100? You could — and indeed increasing speed a little is one way of reducing the effect of a headwind on range as well as speed — but you lose on gasoline, because fuel consumption does not increase in direct proportion to speed. Far from it: the power required and therefore the fuel consumption increases roughly in proportion to the cube of speed. This is why larger engines, while expedient, are an uneconomical way of making an airplane faster.

In practice the cubic proportion between speed and power does not hold true over the entire speed range of a given airplane, because the angle of attack of the airplane varies with speed and the drag varies considerably with the angle of attack. The rule is, however, quite accurate in comparing the upper speeds of airplanes that have appeared with several different engines. The Piper Comanche 180, for instance, cruised at 139 knots; multiplying 139 by the cube root of 250/180 gives 156, which should have been the cruising speed of the Comanche 250: in fact it was 157. The 225-hp Beech Debonair of the same vintage as the

Comanches cruised at 160 knots; its 260 hp counterpart, the P Bonanza, cruised at 169 − which is exactly what the cubic rule would call for. The rule is accurate only in the upper speed ranges, because the rate at which drag decreases with decreasing speed becomes lower and lower as the airplane slows down to a certain speed, below which drag begins to increase as speed is decreased further.

An airplane's total drag consists of two elements. One is induced drag, which is the energy lost in displacing air so as to obtain lift and trim and which may be thought of as the work done in creating a whirling vortex, like a motorboat's wake, behind each wingtip and, to a smaller degree, behind the entire wing. The other is parasite drag, which is all the rest of the drag of the airplane. Induced drag decreases as speed increases, but parasite drag increases with speed. In nearly stalled slow flight induced drag is high, but parasite drag may be only about a tenth of what it is at top speed.

It follows that there is a speed at which total drag is at a minimum, which is neither the lowest nor the highest speed at which the airplane can fly. This unique speed is the speed for best lift/drag ratio (since lift is a constant equal to the airplane's weight); it is therefore the speed for the shallowest glide angle with zero thrust. It would also be the speed for most economical cruise, except that available power increases somewhat with speed, thereby pushing the most economical speed a few knots above the best zero-thrust L/D speed.

Theoretically at least the speed for getting the most air miles per gallon is about 40% above the speed at which the airplane attains its absolute ceiling. Absolute ceiling speed is the speed at which the best-rate-of-climb speed (which decreases with altitude) and the best-angle-of-climb speed (which increases with altitude) are the same. Beech owner's handbooks give this information in graphic form and also give the best-range speed (the official term for the most economical speed); less complete handbooks may give a rule for subtracting or adding so and so many knots per thousand feet of altitude to the best-rate- or best-angle-of-climb speeds; extrapolate to the absolute ceiling, multiply by 1.4, and you have the best-range speed with a few knots. It will be a few knots above the speed for best rate of climb.

This characteristic speed − call it most economical, best range, best cruise L/D, most air miles per gallon, or whatever you like − is determined by airframe characteristics (except insofar as it is modified by propeller-wake effects). Cruising and top speeds, on the other hand, are determined by engine power. Best-range speed is an indicated, not a true, airspeed. The airplane's maximum range is independent of altitude, but it may be covered in the least time by operating at the altitude at which full throttle is required to maintain *indicated* best-range speed. Since best-range power is typically between 40% and 50% and normally aspirated engines produce that power at full throttle at between 12,000 and 15,000 feet, the altitudes providing the best compromise of efficiency and speed are high ones. As a rule, however, the airplane is most efficient if it indicates its best-range speed at any altitude. Best-range speed is used by the very poor, by persons crossing the Atlantic, by airlines, and by pilots who have just switched tanks that their friend told them were full but were found at a quarter. Best-range speed is not to be confused with the speed for best endurance, though one might at a glance be tempted to identify the two. Best-endurance speed is the speed at which the least power and therefore the least fuel is required to remain airborne; it is the speed at which you can remain airborne for the longest time. It is used if you arrive at your destination to find it fogged in, with the fog forecast to lift in late morning or if the tower tells you to expect a delay of two hours for a special VFR clearance because of numerous IFR departures and arrivals or if you have gotten lost over a swamp at night and are flying triangular patterns hoping that some idle controller somewhere will notice you or if you are waiting till dawn to essay a landing with the gear jammed halfway down. Best-endurance speed is nearly the same as the speed for minimum sink in a glide, and it is just below the speed for best rate of climb; since it is the speed at which the least energy is required to sustain flight, it is the speed at which loss of altitude, which sustains flight in a glide, is least; since climb is achieved with energy left over and above that required to maintain level flight, the greatest rate of climb occurs just above (because of propeller effects) the point at which the required sustaining energy is least.

Best-angle-of-climb speed, often called *obstacle-clearance speed*, is usually about halfway between best-rate-of-climb speed and stalling speed. It is lower if the airplane is draggy — with gear and flaps down, for instance — than if it is in clean trim. In general, the cleaner the airplane, the higher its characteristic speeds will be; speeds also go up as the airplane gets heavier.

It is important to be clear about the distinctions between best rate of climb and best angle of climb, best range and best endurance, and minimum sink and flattest glide. The last pair are particularly obscure, because people like to talk about "best-glide" speed — a meaningless expression. If you are gliding down through fog with an iced-out engine and a long prayer to finish, the best glide is the one that gives the slowest sink. On the other hand, if you run out of gas at 7,500 feet about 10 miles short of your destination, the best glide is the one that gives the flattest descent — the most feet of ground covered per foot of altitude lost. Sailplane pilots are well schooled in the difference: minimum-sink speed is used for riding thermals or waves, when you are unconcerned with the distance you are covering but eager to lose as little altitude as possible. Best-glide angle, or best L/D speed, on the other hand, is used for "penetration" — straight legs between areas of lift in which it is essential to cover the greatest horizontal distance with the least loss of height. In fact, high-performance sailplanes carry jettisonable water ballast, which raises the speed at which the best L/D is obtained and makes it possible to reduce the effect of headwinds on a crosscountry trip. The water is dropped if the wind is from behind on a final leg. They can afford the extra weight because the L/D ratio itself — 7 to 10 in a light airplane, over 40 in a good sailplane — is virtually unaffected by weight.

A headwind automatically increases an airplane's best angle of glide (that is, makes it worse) and reduces its maximum cruising range; a small part of the loss may be recouped, however, by increasing speed into the wind. According to a good rule of thumb best-range speed into a headwind is obtained by adding one-quarter of the headwind component (not the actual wind speed but the equivalent direct-headwind speed) to the normal best-range speed. Similarly, the most miles per gallon with a tailwind are achieved by reducing TAS by about one-sixth of the wind component.

To show you how useful all this information is, here is a profile of an imaginary flight on which you would have reason to select several different speeds. You take off from a tree-encircled field with 4 hours' fuel aboard and proceed to a destination 3 hours away. When you arrive, you find the sky partially obscured with one and a quarter in smog, and the tower gives you a delay of 35 minutes. This stretches to 45, and then an unexpected ground fog forms. You are obliged to divert to the nearest VFR alternate, which proves to be 70 nm away. Unluckily, you have a 12-knot headwind on the way. A few miles short of the field you run dry and set up a glide. At 1,000 feet agl, you are still 2 miles from the end of the runway, and it is apparent that you will not make it. The terrain ahead is flat and rocky.

How do you fly it? Initial climb at best-angle-of-climb speed; climb to altitude at best-rate-of-climb speed; cruise at cruise; hold at maximum-endurance speed, reduced by 1 or 2 knots because you are light; proceed to the alternate at best-range speed minus 1 knot for lightness and plus 3 knots for the headwind; initial glide at best L/D speed, given in the handbook and reduced by 1 or 2 knots because you are light; final glide at minimum-sink speed to give you time to turn off all the switches and tighten your shoulder harness, slowing to stall just before impact. You got a pocketful of rocks, but your speed control was impeccable.

26.

MAKING MILES
OUT OF GALLONS

A modern airplane is an efficient transportation device — much more efficient than an automobile, for example. If it is operated at the proper power setting for maximum miles per gallon, a typical single-engine, four-place retractable can go about twice as far on a gallon of gas as can some cars. Even a fast twin such as the Cessna 310Q, with a total of 520 hungry horses to feed, is capable of about 10 mpg, while lighter twins such as the Beech Travel Air and the Piper Seneca can squeeze nearly 14 miles out of a precious gallon of fuel. In terms of passenger miles per gallon light aircraft with reciprocating engines rival even a 747 for efficient use of fuel. Recent CAB operating figures indicate that the Boeing 747-100 quaffs about 3,350 gallons per hour. Assuming a 520-knot cruise speed and a 350-passenger load, the big bird requires 1.6 gallons to transport 1 passenger 100 statute miles. A fully loaded 310Q cruising at maximum-range power needs only 1.69 gallons to carry one passenger 100 miles, and a four-place single needs about 1.4 gallons to do the same task. Lightplanes do an excellent job of moving people and goods efficiently, a fact that should not be overlooked during this time of fuel shortages.

The secret to obtaining maximum fuel economy is knowing at which airspeed to fly. Naturally, proper leaning is essential, and an EGT is probably worth its weight in gold these days. Even the best leaning techniques, however, will not allow you to cope with the fuel shortage if you insist upon cruising at top speed. Typical general aviation aircraft have recommended cruising speeds and cruise-power settings far in excess of what is required for maximum fuel economy. Maximum economy occurs at maximum-range airspeed, which is achieved at a relatively low power setting, usually less than 50%: for a fully loaded Cessna 310Q, for example, maximum range and therefore maximum fuel efficiency occurs at about 45% of rated power. Pulling 75% in a 310Q at 2,500 feet produces about 7.5 mpg, yet at 44% power at 7,500' you can improve on that by nearly 33%. Your cruising speed will only be about 20% slower, a very small inconvenience in relation to the ability to survive in a time of fuel cuts. Cruising the 310Q at 45% power instead of 65% increases miles per gallon nearly 17% at 7,500 feet, and that includes the climb. On a 500-mile trip the slower, more efficient speed adds only about 3 minutes to the trip, and it is quieter and generally more enjoyable.

In theory maximum range — and thus maximum fuel efficiency for propeller aircraft — is achieved in still air at the speed at which the ratio of lift forces to drag forces is maximum. In practice that exact speed is rarely identified as such in the flight handbooks, but it is not absolutely necessary to know it precisely in order to enjoy the benefits of better fuel consumption. It wouldn't be a bad idea, nonetheless, for manufacturers to begin issuing their individual recommendations for most fuel-efficient operation, starting with specific power settings and airspeeds. Flight handbooks usually present range figures at power settings varying from around 40% to 75% If you study these charts, you will be able to identify the power setting that produces the best range. Use that setting but select the lowest rpm that is approved by the manufacturers. Propellers are more efficient at low rpms, because the blades are taking a bigger bite out of the air and the tip speeds are lower.

There are several concepts to keep in mind in attempting to save fuel on a cross-country trip. The speed for maximum fuel economy is not the minimum power required to keep the plane aloft. Minimum-power setting produces maximum endurance, which is fine for holding patterns and flights on which your aim is to prolong the time that you can spend aloft, but the relationship between how much fuel you are burning compared with how fast you are proceeding toward the destination determines best economy. The best fuel-efficiency speed can be influenced by headwinds and tailwinds. You can afford to fly slower and save fuel with a tailwind, but you must fly faster into a headwind. To illustrate this point, consider flying into a headwind whose velocity is equal to the zero-wind maximum-range speed. No progress would be made toward the destination: miles per gallon would equal zero. But if you flew faster, you would at least be getting somewhere, and the miles per gallon, of course, would be greater than zero. As a rough rule of thumb in flying into a headwind, add enough extra power to increase the recommended maximum-range speed by an amount equal to one quarter of the headwind component; in flying with a tailwind reduce power sufficiently to fly at the maximum-range speed less one-sixth the tailwind component.

The indicated speed for maximum range does not change much with altitude, but it is altered slightly by weight. For a 5,000-pound light twin the indicated airspeed for maximum range should be reduced about 1 % for each 100 pounds that the plane is under gross; for a 2,500-pound single reduce the recommended power so that the maximum range IAS is slowed by about 2% for each 100 pounds under gross. As the plane burns off fuel, the indicated airspeed should be reduced accordingly for maximum mpg. Fuel efficiency will be better if the plane is light, so don't carry around any unnecessary weight. From the viewpoint of getting more passenger-miles per gallon, however, make the most of each trip by using all available seats.

In practical terms fuel efficiency is not affected by altitude. What slight benefit there is in flying higher in an unsupercharged aircraft is generally negated by the fuel and time required for the climb. Cruise altitudes should be selected on the basis of winds aloft and anticipated duration at the cruise altitude. If there is a headwind, stay as low as possible to minimize the time spent at high fuel-flow rates needed for the climb and to maximize the amount of the flight at lower altitudes, where the headwinds should be less. Remember to increase your power setting just enough to add the velocity of the headwind component to the best-range indicated airspeed. If there is a tailwind at altitude, go after it, provided that you plan to ride it long enough to overcome the extra time and fuel needed for the climb. Climb time should be kept to a minimum: in theory best-rate-of-climb airspeed should be used, but in practice not much is lost by cruise-climbing to altitude, and forward visibility is considerably better at cruise-climb angle of attack. In the interest of safety use factory- recommended cruise-climb numbers: it costs little in fuel efficiency. Select your cruising altitude by considering (1) the difference between indicated cruise airspeed and climb airspeed, (2) the duration of the climb, and (3) the ratio of the climb fuel flow to the cruise fuel flow. Multiply all three factors and divide by the amount of tailwind available at the selected altitude. (Remember to use knots for airspeed if you use knots for the wind.) The result is the number of minutes that you should remain at altitude to gain back the extra fuel used to climb there in the first place. If it is a short trip, there is no point in wasting a lot of time and fuel to climb up after it.

Consider a Cessna 310Q climbing at 120 knots IAS and burning 28 gph. Taking off at gross weight, the plane could be at 10,000 feet in about 12 minutes, and it would have consumed about 6 gallons of fuel for the climb. Once it was established at a 45% power cruise, the 310 would be indicating 128 knots and would be sipping 17.8 gph. The product of airspeed difference (8 knots), duration of climb (12 minutes), and ratio of climb fuel flow to cruise fuel flow (27 divided by 17.8) is 8 x 12 x 1.57, or 151. With a 10-knot greater tailwind at 10,000 feet than at sea level, therefore, the pilot needs to spend about 15 minutes at altitude to make up for the extra fuel that he used for the climb. For this example even a slight tailwind aloft is worth pursuing, but for another aircraft with a larger difference between climb and cruise airspeed and a lower rate of climb (the fuel-flow ratio is usually about 1.5 and doesn't differ much between planes) the numbers might indicate that it is better to accept a 10-knot nudge at 3,500 feet than to struggle up to 10,000 for a 25-knot push.

For the letdown it is better for fuel efficiency to remain at the selected cruise airspeed and to reduce power for the descent. The most efficient speed is determined mainly by the aircraft's aerodynamics; attempting to use more power than is needed to hold that speed is technically inefficient. Plan your letdown so that rpm and manifold pressure stay within safe operating limits, assuring proper cylinder head temperatures, and start the descent so that you are at pattern altitude close to the destination airport. The power that you don't have to add at the end of the ride adds to the total fuel efficiency. Reducing manifold pressure 5" results in a 500-fpm rate of descent at cruise airspeed, and figuring 2 minutes for each 1,000 feet of descent makes it easy to calculate the point at which to start down, provided that you have some notion of your groundspeed.

Whenever possible, fly the most direct route VFR. IFR flights require 10% to 15% more air time, which takes a big bite out of fuel efficiency. If IFR is necessary, telephone the tower − or radio them before start-up − to confirm that they have your clearance and that there will be no delays on the ramp. Remember: anything you do to minimize unproductive engine time saves fuel and adds to the efficiency of the lightplane.

27.

THE MOST

DANGEROUS MYTH

A couple of million student pilots ago most people were sure that airplanes would fall out of the sky unless they were held up by incredible skill, brute force, and the pilot's seat belt. In time this led flight instructors to begin saying things such as: "Look, dummy, this airplane wants to fly! If you'll let it alone, you'll find that it can fly better than you can!" This was valuable − but only to encourage students to relax and to be more gentle in their use of the controls. The truth took too long to tell, and there is reason to doubt that the instructors of those days even knew what the truth was.

After a while all students were exposed to what became known as "confidence-building maneuvers." Once the idea became accepted as good medicine, instructors and others who ought to have known better began giving massive doses of it. Like overdoses of many drugs, the effects were sometimes fatal. Probably the best known of these medications comes from *Civil Aeronautics Bulletin 32*, issued years ago by the Civil Aeronautics Authority and still being reprinted, read, and believed by instructors and students alike. The most outrageous claim is this: *"The plane will recover from a bank and turn by itself. This will be demonstrated (by the instructor) by putting the plane in a medium bank and releasing all the controls. The nose will immediately drop a little, and the plane will slip toward the low wing, but after a short time will return to level flight."*

This statement is not true. A section of a report done on an FAA contract by Cornell Aeronautical Laboratory comes close to the truth "The disorientation of non-instrument-rated pilots upon encountering IFR or marginal VFR conditions each year forms a large percentage of general aviation accidents. The basic cause for this disorientation is a lack of spiral stability in the majority of general aviation airplanes. This lack of stability means that the tendency for the airplane to maintain a constant wings-level attitude, if left unattended, is at best marginal. If the airplane is spirally unstable, it will, if uncorrected, enter a turn in which the angle of attack slowly increases and the nose slowly drops, with a resultant increase in airspeed as the turn develops.

"Once the spiral develops, the airplane will eventually fly into the ground in a spiral dive if the pilot . . . does not make suitable corrections."

It is hard to guess why Cornell used such gentle language. The complete truth is that even if an airplane has been built precisely as it was designed and rigged precisely as intended and loaded so that the center of gravity is exactly in the right place − both fore and aft and side to side − it will not fly wings-level, hands-off for very long. Something will disturb the delicate balance, and a wing will go down; once the wing goes past a critical angle, it will keep on going down, and a spiral dive will develop.

As long as the pilot retains visual contact with the real world outside the cockpit, he has plenty of time to pick up the low wing, If the pilot loses outside visual reference, bad trouble follows if he is not instrument-trained. Without visual reference man has an inherent spiral tendency. This was discovered and proved by Major C. Ocker at Crissy Field in 1926 and further explored in later tests by Major Ocker and Lieutenant Carl J. Crane at Kelly Field. Pilots were blindfolded and then required to walk or drive a car or steer a boat in a straight line. Every time the resulting ground path was a clock-spring-spiral. Not long afterward Lieutenant Crane found the same tendency in pigeons. They were blindfolded and released from an airplane, and the result was reported in the famous *Blind Flight in Theory and Practice*, by Ocker and Crane: "The pigeons without exception performed all manner of evolutions which indicated lack of flight control, including stalls and spiral dives. Finally holding their wings at a high dihedral angle, they descended to the ground in much the same manner as a parachute."

The result is a perilous man-machine relationship: both partners − man and airplane − have identical spiral tendencies, both of which depend on the pilot's view of the real world. Neither one can recover unaided, and the most common form of pilot "aid" is instrument training.

Stability and controllability can be said to be opposed, and a balance between them must be chosen on the basis of the mission that the aircraft is to perform. The result is often called the handling qualities of the aircraft. Handling qualities considered excellent for a military fighter would not be acceptable in a heavy transport, just as the handling qualities of a competition aerobatic airplane would not be suitable for a four-place family airplane.

The reason why a compromise is necessary is that it is at least theoretically possible to build an aircraft so stable that it would not be controllable at all; in fact it might not even move. At the other extreme, controllability can be emphasized so much that the resulting low stability level would in turn affect controllability in the wrong way. If this seems odd, it is a fact that there are many large airplanes and military fighters now flying that depend on full-time stability augmentation and are not safe without it. Details differ, but in general all of the design-decisions concerned with handling qualities follow the same pattern. Directional stability, which involves pitch and yaw, is relatively high, or stiff; roll stability, which involves turning, is relatively low, or soft. All three stability elements work quite well together while the airplane's wings are level or nearly so in the sense of keeping the machine going along straight and level at a fairly steady height and speed. If a wing goes down, though, things change for the worse.

Yaw is not really a control axis: the function of the vertical pieces on the tail is simply to keep the tail following the nose. Because certain flight conditions demand more power in the yaw axis than the vertical pieces can supply if they are fixed, a movable section − the rudder − can be used to magnify the yaw effect. Rudder is used to overcome adverse aileron yaw while entering or recovering from turns. Under conditions of high angle of attack and high power, as in climbs and slow flight, right rudder is also needed to keep the airplane from turning left;

if power and the nose are low, as in glides, a touch of left rudder is usually required to keep the machine from turning right. In conventional twins the inadequacy of the fixed fin becomes very clear if one engine stops, and a great deal of rudder must be used. As the Ercoupe showed, the rudder can be eliminated as a pilot-operated control. Several other airplanes come close to this by mechanical interconnection of ailerons and rudder − another of the clever ideas of Wilbur and Orville Wright.

The pitch axis is unique to aircraft and submarines, and, since pitch is the axis in which planes challenge gravity, pitch stability must be firm. It takes more effort to move the controls in pitch because of this, but, because pitch-control requirements are primarily modest and not particularly critical as to timing, this is acceptable. (In landings, pitch control is time-critical, which is one reason why it takes so long to perfect landing skills.) The standard pitch-stability demonstration involves setting power and pitch trim for level flight, then applying back pressure to slow the airplane by 10 mph. The stick is then returned to neutral (and students who watch closely will often observe that the instructor pins it there, because "stick-fixed" stability is higher than "stick-free"). The nose comes down, and the airspeed increases. The increase in speed produces an increase in lift, which starts the nose up again. As the airplane begins to climb, the speed begins to drop, and finally the nose will start down again. In most light airplanes this slow vertical dance − called a *phugoid cycle* − stops after a couple of ups-and-downs, leaving the airplane at the same altitude and airspeed as at the beginning. If the vertical oscillations of the flight path in this exercise should get wilder instead of milder, the plane is said to be divergent in pitch, and something must be done about that right away. Modern lightplanes have good pitch stability characteristics, though some are far from ideal when the center of gravity gets close to the aft limit.

The roll axis is the one in which you need the greatest controllability, because maneuvering must be done more quickly and more precisely. Stability in roll is therefore subordinate to controllability. This is where things start to get sticky, so you should take a look at some stability fundamentals. A pendulum, if displaced, will positively return itself to its starting position. A billiard ball on a level table is neutrally stable: it will come to rest somewhere after having been moved, but it doesn't care where it comes to rest and has no tendency to return to the place that it started from. A pencil will not stand on its point and is thus unstable: if you let it go, it will fall. It will not stay put, and it will not even try to get itself upright again. Almost all airplanes reflect all three conditions of stability in roll. For very low angles of bank, they tend to return to wings level. At a narrow range of medium bank angles, they tend to stay where they are put. At higher angles of bank, they tend to increase the bank angle. The first two conditions seldom last very long if the controls are released, but they can easily be observed. The self-righting tendency is what makes gentle turns so hard to do well, because the airplane keeps trying to shallow the turn; the staying-put tendency is taken advantage of in instrument work, since it commonly occurs at the bank angles associated with standard-rate turns. A perfectly built, perfectly rigged airplane, with its center of gravity exactly at the lateral center, is little more than a scientific concept. Even if everything is right − at the beginning of the flight, you can bet that the fuel load will soon become unbalanced if nothing else does. It doesn't take much sideways-weight bias to set the airplane up for a spiral dive, and the most probable trigger will be a gust, even a small one.

The yaw stability of the airplane comes into play here. It still does its job of keeping the tail following the nose, but, as it does so, it forces the up wing to fly faster, which increases its lift still more, and the bank angle increases even more. Pitch stability makes it own contribution. As the nose comes down and the speed increases, the effect of pitch stability is to increase the angle of attack, just as it does in the wings-level phugoid. But since it has no way of knowing that a wing is down, the result is quite different. A simple flight experiment will give a graphic demonstration of what happens.

At a normal high-airwork practice altitude, establish and stabilize a normal turn with a bank angle of 50 degrees or so. Without changing anything else add back pressure. Watch the low wing as you do this, and you will see that increased elevator pressure causes the airplane to roll tighter into the turn. Why? Although back pressure applies the same angular change to both wings, the high wing will benefit more, because it is going faster than the low wing. This is also why increased back pressure won't fix altitude loss in steep-turn practice. The fixed horizontal tall surface does the same thing, though not so rapidly. The effects — the fin forcing the high wing to fly faster and the differential-lift effect of the horizontal stabilizer — are both in the same direction: both tend to tighten the spiral, and, as the spiral tightens, the speed increases.

It ought to be easy now to see why the pilot's normal, trained reaction of trying to reduce airspeed by back pressure on the yoke is dead wrong. At best this will simply tighten the spiral still more; at worst it will produce stresses high enough to break the airplane. This is the root cause of the characteristic signature of an accident caused by an untrained pilot losing control in instrument weather: the airplane is seen spinning out of the clouds with parts already separated from it. It also should be clear why it is that pilots taking instrument training are firmly taught to stop the turn first before trying to recover from the dive. Finally, it ought to be clear that the statements that the airplane wants to fly straight and level, that it will do so indefinitely by itself, and that it will recover by itself to straight-and-level flight from a turn if only the ham-handed student will turn it loose are simply not true.

A pitifully small number of practitioners of the aeronautical arts have thoroughly understood this problem for at least half a century. One of the most thorough and readable explanations appears in Wolfgang Langewiesche's classic *Stick and Rudder* in the chapter titled "What The Airplane Wants to Do." The only serious fault to find with it lies in the author's omission of the connection between spiral tendencies in man and machine.

In the middle 1920s, the old National Advisory Committee for Aeronautics issued a paper on a simple way to keep the wings level. This required only the standard-rate gyro used in the turn-and-bank indicator, with its axis tilted 45 degrees so that it could sense both yaw and roll, coupled to the ailerons. This idea lay unused until 1940 when DeFlorez adapted it for use with target drones. The late Alick Clarkson took the idea a major step further, added means of control, and turned out a simple kind of autopilot, which eventually became the Brittain.

In the 1965 model year one light-airplane manufacturer took a bold step: using the name *Positive Control* (PC), Mooney equipped all their production airplanes with a wing-leveler made by Brittain. There were those who said that it was a gimmick and that it would do no good. NASA, NACA's successor, was the first to prove them wrong. Norman Driscoll of the NASA Langley Research Center presented a paper at the Society of Automotive Engineers Business Aircraft Conference in Wichita in 1966 titled *The Effect of a Light-aircraft Stability − augmentation System on Pilot Performance*. Its summary states: "A simple wing-leveling device for light aircraft, and the effect of this device on the instrument-flight performance of non-instrument-qualified private pilots, were evaluated.... These private pilots were provided with an increased capability to recover from an inadvertent instrument-flight situation. Limited navigation communications tasks, beyond the capability of the pilot while controlling the basic aircraft, could be performed with system aid. *Control wheel force developed by the system as a function of bank angle in a steady turn was found to be a definite instrument flight aid.*" That last sentence has an important bearing on the acceptance of PC, as we shall see.

The Langley paper was vulnerable to scientific criticism in one respect: only two pilots were tested. The Cornell Aeronautical Laboratory study later the same year disposed of that. Cornell tested 26 nonrated pilots and compared the result with tests of 5 experienced, rated pilots. The Cornell study was sponsored by the FAA and is called *Flight Evaluation of a Stability-augmentation System for Light Airplanes*. The equipment installed was the practical equivalent of a Brittain autopilot system and therefore included pitch stabilization. (One of the curiosities of the project is that the FAA chose to allow an equipment-development contract to build the Brittain equivalent at a considerably higher cost than they would have had to pay for an off-the-shelf Brittain.) The Cornell summary of the non-rated-pilot results made 10 conclusions, of which these are the most significant:

● Performance during VFR flight is invariably better than during IFR flight.

● In VFR flight whether the SAS (stability-augmentation system) is on or off generally does not make much difference in pilot performance.

● In IFR flight, having the SAS on invariably improves the pilot performance.

● With the SAS on, the difference between VFR and IFR flying generally disappears.

There is one more significant report: a 1970 study by the Defense Systems Division of Bunker-Ramo Corporation, funded jointly by the FAA and the Air Force Flight-dynamics Laboratory. Its title is *Flight Evaluation of a Pilot-assist Stability-augmentation System for Light Aircraft*. In contrast to the Langley report's comment that wheel forces proportional to bank angle were desirable, some of the pilots involved in the Cornell study had strong feelings against the force required to hold the airplane in a steady turn. The Bunker-Ramo project, recognizing these complaints, sought to find a way to preserve the value of a wing-leveler without introducing wheel forces that pilots wouldn't like.

The scheme was to make a standard Brittain wing-leveler, like the ones in the Mooney PC (and in a great many other airplanes as well), but to limit authority up to 15 degrees of bank. You will recall that this is in the neutral-stability area of a standard-rate instrument turn. If the airplane − or the pilot − tried to exceed this bank angle, the wing-leveler came on at full power, which Bunker-Ramo called a "wall of force." Unfortunately, the equipment didn't work as it was supposed to, and the results from that part of the study are uncertain. Even so, the results again made it clear that some sort of simple stability-augmentation system applied to the roll axis alone makes IFR flight about the same as VFR flight for nonrated pilots.

To those few who have long understood the problem of an airplane's inherent instability these three research projects must look like cracking walnuts with a cannon: the rest of the aviation community has received the findings with magnificent unconcern. Yet the objections about continuous pressure to hold the airplane in a turn are respectable and must be dealt with.

There is an even more respectable objection to accessory appliances, for there is nowhere in the world any instrument, accessory system, or piece of avionics gear that has demonstrated the same reliability as a basic airframe. The standard-rate gyro, driven by air, electricity, or both — the heart of nearly all wing-levelers — comes pretty close, but nobody who heads a company that makes instruments, accessory systems, or avionics seems to have ever taken the basic decision that they must be made to reach the reliability level of the airframe. These things are more delicate than sheetmetal and their health is not so easily measurable, but these differences must be accepted as challenges; someday somebody is going to meet them.

Even though one can be optimistic about the eventual success of this approach, it is in an important sense off the mark. There is only one airplane that has positive roll stability built into the airframe itself — Moulton Taylor's Aerocar, which has a design type certificate from the FAA but is still a long way from production. You can roll the Aerocar into a turn of 45 degrees of bank or more, release the controls, and it will roll itself out wings level. Taylor himself seems to be less impressed with this achievement than most people. He says that anybody who has ever designed, built, and successfully flown a radio-controlled model airplane has had to find the same solution.

Taylor's explanation is shockingly simple: the roll stability of any airplane depends on the exact geometry of roll axis. In the ground-school diagrams, the line of the roll axis passes nearly through the physical centerline of the airplane from nose to tail. In the very early stages this error may be excused, but because it is almost never subject to later refinement, it leaves a hole in the student's understanding.

The truth is that in almost all airplanes, the roll axis inclines downward from tail to nose. This is the basic cause of the yaw stability forcing the up wing to go still more up and the down wing to go still more down. In the Aerocar, the roll axis points upward from tail to nose, and the result is that yaw stability reinforces the dihedral-stability effect of the wings. To simplify: the physical location of the fin determines the inclination of the roll axis. If the fin sticks up above the fuselage, the roll axis points down. The Aerocar fin sticks down below the fuselage, so its roll axis is pointed up. Not many model years ago Cessna made a striking improvement in low-speed roll stability of the 310 series by adding a ventral fin (below the fuselage); the F-4 jet fighter has achieved an important degree of improvement in roll stability by bending the horizontal tail pieces downward in an inverted V. Some other high-performance fighters have chosen the same solution.

There are two more major points of discussion about the auxiliary, bolted-on wing-leveler, one technical and one philosophical. Both involve the subject of the wheel force required to overpower the system.

As we have seen, the Langley study found wheel force desirable without qualification, but the pilot sample was small. The Cornell contract found considerable resistance: in addition to the complaints that holding constant pressure to keep a turn going was awkward a few pilots claimed that this also made holding pitch attitude harder. A smaller number actually forced the airplane against the restoring force to the point at which a disaster might have occurred if there had been no safety pilot to take over control. In addition to the Bunker-Ramo "wall of force" concept there was another attempt to save the principle while overcoming the objections. This was conceived by Dr. Karl Frudenfeld, who was president of Brittain Industries during its growth phase and who had a great deal to do with the Mooney PC. It was a simple system and worked well, though it has not been exposed to any large group of pilots. (The reasons that it never got to market were commercial, not technical.)

First, the system was proportioned so that the wheel force needed to overpower the system was much less than in the PC. Second, the authority of the system was related to airspeed. This was done by using a large fiberglass venturi, nestled inside the cowl in an ice-protected area, as the vacuum source instead of an engine-driven vacuum pump. As the pilot slowed during the landing approach, the servo power was reduced, and, by the time he was ready to flare, it was insignificant. At the other end of the speed range the wing-leveler's authority increased with airspeed. The gyro was not affected, because all Brittain gyros in recent years have been redundant in the sense that they are operated by both electricity and vacuum.

On the philosophical (or scientific, if you prefer) side, it seems that the preoccupation with eliminating pilot objections misses an important point. All the studies contain a serious bias: every subject was a licensed pilot, with all the trained responses that every pilot has to learn. That approach was expedient, but it leaves unanswered the larger question of how nonpilots would have responded. To resolve this, two groups of student pilots, all with no previous flight training, should be compared after the same amount of training — one group in standard airplanes, the other in airplanes with stability augmentation.

In the world's most common travel machine, the automobile, there is a direct connection between steering-wheel force and rate of turn so that everyone who drives a car knows from the beginning that the turn will stop if he relaxes pressure on the steering wheel. The fact that an airplane acts differently doesn't necessarily mean that it acts better: it would seem obvious that people who don't already know how an airplane acts would welcome a control response like one with which they are already familiar. Probably most of the present pilot population will accept it too, once they understand what it will do for them and their passengers.

Even though none of the studies found any significant improvement in VFR flight with the SAS, it can be of considerable value to VFR. It reduces fatigue on long flights, and it holds reasonably close to a heading if the pilot has to read a chart or do other bookwork. Under IFR conditions, it is no less than essential.

Considering the money that the FAA has spent on two of the projects cited, an optimist might think that the observation of positive results in safety would persuade the FAA to belly up to the problem and insist that positive roll stability, by whatever means of complete reliability, become a condition of certification of all airplanes (with some exceptions, such as aerobatic models). There are no signs that this is imminent nor any signs that other makers will follow the Mooney lead.

Until positive roll stability is more widely available, whether by regulation, manufacturers' decisions, or individual pilot choice, there are two things every flight instructor should do: First, he should forget *Civil Aeronautics Bulletin 32* and all its descendants and stop telling people that airplanes fly very well by themselves and that they will return themselves to level flight out of a medium turn. Second, and even more important, he should demonstrate the airplane's spiral tendency and let the spiral dive develop far enough so that there is no doubt in the student captain's mind about the eventual outcome. After demonstrating this VFR the exercise should be repeated with the student under the hood.

Bad as they are, the fatal accidents may well be a smaller part of the problem than are the number of pilots who may have given up flying after no more than one shattering exposure to a spiral dive − with no accident − after having been solemnly assured throughout their VFR flight training that the airplane knows how to fly level by itself.

28.

ATTITUDE + POWER = PERFORMANCE

Before us is an old piloting proposition: throttle controls altitude and elevator controls airspeed. Also before us is a letter emblazoned with the official emblem of the Federal Aviation Administration and containing these words: "Power controls *speed* and elevator controls *altitude.*" We must weigh these contradictory claims and decide which school of thought bears the truth. It helps some to know that in 1976, the FAA reversed itself and adopted the throttle/altitude, elevator/airspeed position, but that didn't happen until an enormous amount of debate had taken place. For all we know, they may change their minds again, so this controversy is no mere scholastic exercise. Let us descend, then, hand in hand, into the pit of sticky verbiage and viscous substance to see where the "truth" really lies − or flies.

The concept of using elevator to control an aircraft's speed has its origins in aviation's hazy past. Somehow, somewhere − perhaps under the wing of a biplane in some farmer's field or in the back room of a World War I barracks − the idea of connecting the action of the elevator to the indications on the airspeed gauge first evolved into a recognizable form. Until well after World War I most pilots controlled airspeed with the elevator; altitude, with throttle; and used needle, ball, and airspeed to stay upright. Pilots slowly began to discover a new instrument: the artificial horizon. The military flight-training instructors were the first to make this discovery, although even they had allowed years to slide by before realizing the advantages of teaching pilots to control flight by attitude rather than by performance gauges. The concept of attitude instrument flying was born, with its litany "attitude plus power equals performance" as the crux of the new teaching experience. The which-controls-what relationship was soon to be reversed.

Many of the participants of the era say, "What we had been doing with our airplanes forced us to change the theory to match reality. If you watch closely, you soon discover that pilots control their altitude with the elevator, not with the throttle." In an FAA flight-instructor revalidation course, one would hear the same thing. "The elevator controls not speed but altitude," the government instructors would preach to scores of pilots. Some pilots would walk away as converts, while others would absolutely reject what they'd heard. Many could not decide. Was the FAA attempting to legislate a change in the laws of physics, or were many airmen simply following a procedure that was both a day late and a dollar short?

The truth is that neither the elevator nor the throttle independently controls

143

airspeed or altitude. Moving the throttle simply creates more or less thrust: the elevator merely adjusts forces so that the angle with which the wing bites the air can be changed at the pilot's will. When an aircraft's lift exceeds its weight, it gains altitude. Greater lift can be obtained by increasing either the airspeed or the wing's angle of attack. An increase in airspeed can follow if one pushes in the throttle or lowers the nose. About the only completely accurate statement that can be made about the use of cockpit controls is simply that the elevator controls the attitude of the aircraft and that the throttle controls the power. If this sounds familiar, look back to the concept of attitude instrument flying for the clue: attitude (elevator control) plus power (throttle control) equals performance. Nowhere in the attitude-flying concept is there a mention of which aspect of performance is controlled by which handle. The intention of the idea is obviously to show that elevator and power combine to create performance: pilots from either of the two schools who argue otherwise are simply wrong. Even the much-touted FAA *Instrument Flying Handbook* is quite specific: "The pitch instruments are (among others) the altimeter and the airspeed indicator.... The altimeter gives you an indication of pitch in level flight.... The airspeed indicator presents an indirect indication of pitch attitude.... In this handbook, the altimeter is normally considered the primary pitch instrument during level flight."

If nowhere in the body of attitude-instrument-flying doctrine is it specifically alleged that specific aspects of performance are tied to specific individual controls, then why did the need arise to postulate such a connection in the first place? One reason was to provide a handy rule that could be used to expedite pilot training and guarantee a workable level of safety from pilot actions in various situations. Armed with their rule, claim the advocates of each school, a new pilot can be trained more quickly and effectively and an old pilot made to fly more safely.

The FAA's contention that we should think of elevator as the altitude controller had advantages in the sphere of training. One can present a logical case to show that 99% of the times that a pilot pulls or pushes the yoke, his intention is to capture a certain altitude or rate of change. Think about a typical flight, and the truth in what the FAA was saying seems obvious. Cruise flight and instrument approaches are two examples of occasions when the elevator is apparently used to manipulate altitude while throttle controls the airspeed. When ATC requests a flight to slow down, the pilot invariably comes back on the throttle, not on the wheel, further verifying the FAA's observation that "this is how we really fly, so it's the best method to teach."

The elevator-controls-airspeed people could then be heard rising to the battle: "What about on takeoffs or when the engine quits − or in a glider?" The FAA then patched up their shot-through blanket with a qualification: "The concept works whenever power is both *variable* and *available* for the purpose of controlling speed. That covers most of our flying time, therefore making it the better concept to teach. Naturally, if the power is not variable and available, then the pilot must use the elevator to adjust his speed."

"How about spins?" asked the elevator-controls-airspeed group. "Well, that's not a normal situation, and our concept is used primarily as an aid to teaching." That sounded like qualification number two. "See? Controlling air speed with the elevator's safer because it always works whenever there's any altitude left." "But we don't really fly by controlling airspeed with the elevator, so why teach it?"

"Because it's safer to use!" "But it's harder to teach!"

Both groups were wrong. Their error lay in the premise that there is a direct connection between elevator and throttle, on the one hand, and airspeed and altitude, on the other. They erred because they had not thought about how pilots really fly. Pilots control either airspeed *or* altitude with the elevator. The choice depends on which facet they're more concerned with at the moment. Analyze a pilot at work, and you'll see for yourself. During takeoff he cares about gaining altitude, but he is more concerned about maintaining the proper airspeed − and the elevator assumes that important task. During cruise the main concern becomes one of maintaining a constant altitude, and it is the role of the elevator to satisfy the new need. Normal descents are usually executed by reducing the power to an acceptable point, then flying a specific airspeed or rate of descent, using the elevator to achieve either of these goals. While the major concern of the pilot may alternate between airspeed and altitude, the one fact that remains constant is his manipulation of the elevator to accomplish these tasks. The FAA analysis that pilots spend 99% of their time controlling altitude with the elevator is true simply because the real world of aviation is more precise about altitudes than airspeeds. We fly at *precisely* 8,000 feet but we accept whatever forward progress we can get.

Flying is actually accomplished by the concerted use of attitude (elevator action) plus power (throttle action). The best device for controlling the forces of flight is the elevator, because power is so difficult to adjust precisely. This hypothesis addresses another myth, this one surrounding flight control in turbojets. Since the throttle of a jet engine is less positive in its effect on flight (due to spool-up and lack of propwash), pilots find themselves adjusting the elevator more often and the power less.

The FAA is correct when it says that people are best trained if they learn the methods that they'll really use. Provide a system that is both safe and effective, though, and everyone will be covered all the time. Rather than confusing the issue with synthetic techniques both schools would be better off abandoning old lines and stressing the sole fact that stands out loud and clear: the throttle is an important adjunct, but the real business of flying is done with the yoke.

By teaching pilots to control their most pressing performance requirements with the elevator they are directed to the most sensitive, accurate, and dependable device in the cockpit, a device that can take care of their greatest momentary need. Stall or spin recovery, cruise flight or flying the glideslope to a bare-minimums landing all are jobs that belong in the control department − which quite literally places those tasks directly in the hands of the pilot.

29.

ILL WINDS...

When winter winds howl, airplanes get bent. This is not only unfortunate but often avoidable, since even the smallest of our birds do well in strong surface winds when handled carefully. A tricycle-gear airplane can tip over if it is mishandled, though − even in a moderate wind − although the airplane would have remained upright and intact had the pilot only practiced the basic principles of taxiing in wind. The airplane manual often contains a chart showing the proper control positions to use in taxiing with the wind from each quarter. For example, if the wind is from your right rear, the aileron control should be to the left in order to position the ailerons to counteract the wind's tendency to lift the right wing. There is usually also a warning to taxi slowly and to make turns from a downwind taxi with extra care. And to make that run-up headed into the wind. Low-wing airplanes can tip too − it is not a sport reserved for high-wing airplanes − so the rules apply equally to all.

What is the maximum wind velocity for light-airplane taxiing? The answers to this question are as variable as the kinds of wind. At some airports they lash the airplanes to the ground and flee to the tavern when the wind reaches 25 knots. In Kansas where the wind really blows, 25 knots is considered a light breeze. Most light airplanes begin to get very restless on the ground when the peak gusts reach 35 knots, though some of the heavier craft can handle 50 knots as long as there's a wind-wise pilot at the controls. Ground operations are possible in a Skylane at 45 knots, but taxiing has to be done very *slowly* and carefully, and it is a real relief to finish the day's business and put the airplane away. It is sometimes necessary for a wingman to hold each wingtip down until you get into takeoff position.

Wind should pose no great takeoff or landing problem if it is within taxi limits and blowing down the runway. Crosswinds, however, demand effort. The most common cause of botched-up crosswind takeoffs is premature liftoff. Any difficult chore makes the time go more slowly, and a pilot charging down the runway in a strong crosswind is eager for the takeoff roll to end and the flying to begin. If he's having any problem with drift or directional control, the tendency is to try to eliminate the problem by flying away. As the nose is lifted and the wing begins creating lift and taking weight off the wheels, the crosswind asserts itself by drifting the airplane sideways on the runway. The scuffing of the tires makes the pilot want to be airborne even more, and, instead of doing what he should do (applying control to get rid of the drift), the pilot lifts the nose even more. The airplane then flies and begins drifting more rapidly once the wheels are free of the runway. If at this point the airplane settles or stalls because of a sudden drop in wind velocity or simply because it flies out of the ground cushion and comes back to the ground with a lot of drift, that's where the old saying about rolling it up in a ball comes from. Heed the textbook when it says to accelerate to a slightly higher speed than normal before lifting off in a crosswind − to assure that, when the airplane begins to fly, it does so positively.

One other takeoff item: if the best possible climb is needed after a crosswind takeoff, any turn should be into the wind (obstructions permitting). Why? Is it not true that the airplane operates in the air without regard to the wind? Yes, but there are exceptions. Wind usually increases in strength with altitude, and this can be very pronounced at lower altitudes. When you are climbing *into* a wind that is rapidly increasing in strength with altitude, you can get quite a boost out of it. When you are climbing *with* a wind that is increasing in strength, you can get quite a sag out of it. If your takeoff is in hilly country, it's not a bad idea to study the terrain and to visualize where the updrafts and downdrafts will be so that you can stay away from the obvious downdraft areas. (Updrafts are on the upwind side of hills, downdrafts on the downwind side.)

Crosswind-landing problems are created by three basic mistakes: fouled-up traffic patterns, excessive speed on approach, and plain old white knuckles. The worst traffic-pattern problems are created by strong tailwinds on the base leg. In fact, many pilots come to grief before they even begin their final approach because of a strong base-leg tailwind: they overshoot the turn onto final because of their high groundspeed, which in turn gives them the visual illusion of adequate airspeed. They then tighten the turn to get back in line with the runway, and a stall/spin is the result. The preventive technique is to fly a very wide base leg if there is a tailwind on base. It doesn't hurt to plan a slightly longer final approach either.

On final the windy-day approach speed should be the speed that you normally use plus the value of the gusts over the steady wind speed (10 mph extra if the wind is 15 gusting to 25, for example). Too much speed on approach results in too much speed to dissipate before landing, which leads to the temptation to try to force the airplane onto the runway. Wheelbarrowing and loss of control often follow a supersonic landing, and busted nosewheels, props, and wingtips are the end result. A tricycle-gear airplane simply is not going to be manageable on the runway unless it gets there with some weight on the back wheels, which it does not have after a level landing at high speed. A tail-low touchdown is usually easier to accomplish if you leave the flaps up. This dictates a higher approach speed than with the flaps down plus the extra for the gusts, but don't overdo it. And remember that you'll need a longer runway than usual.

White knuckles? If a pilot can relax a bit somewhere out there in the pattern and ungrip the wheel slightly to let a bit of color back into his knuckles, the approach and landing will go better. Airplanes were made to fly, not to fight, and the windy-day approach and landing should not turn into a wrestling match.

Whether your knuckles are red, white, or blue, remember that surface obstructions around the airport can create extra-tricky conditions if the wind is up. Hangars or trees west of a north-south runway, for instance, can disturb a strong westerly wind enough so that you may not be sure of how it will affect your airplane when it reaches the runway. If you case the airport from altitude first and relate the position of the obstacles to the indication of the wind sock, you can usually anticipate any real surprises. On a long runway it is sometimes possible even to beat the wind by coming in high and touching down past what looks like the point of maximum disturbance.

Strong winds can also be a factor en route — especially if you are flying over rough terrain. As the wind blows over such a surface, it causes turbulence that can really beat you up. This vertical variety of terrain and the strength of the wind determine the severity of the turbulence, with the worst found downwind of the rough terrain. Climbing above turbulence always seems as though it will be the best bet, though it can be a long climb, and, if you're headed upwind, you'll probably find the headwind stronger aloft than down low. The time passes faster in smooth air, though, so climbing above the bumps is usually a good investment.

If you elect to stay down low, the ride would become extremely uncomfortable before the turbulence would actually menace the structure of the airplane. (There is an occasional case of sudden structural damage in unexpected turbulent air, however — usually when severe turbulence is forecast in the vicinity of mountains because of high winds flowing over the rough terrain.) If you slow down, you'll be helping not only your comfort but your airframe: a lower airspeed means less strain. Maneuvering speed is the speed of maximum strength for your airplane: if you want to fly with the best structural advantage, fly at or slightly below that speed. And remember that your best speed in turbulence *decreases* with weight, because the lighter the airplane, the more bang from the bump. If the load is light, slow it down even more. If you are cruising in smooth air, be careful of descending by just pushing the nose over and letting airspeed build up as you let down. Somewhere during that descent you might run into some bumps, and you don't want to hit them with the airspeed up in the yellow arc. (That yellow arc is aptly colored: when the airspeed gets into the yellow, get chicken and slow down.) If ordinary turbulence is suspected, have the airspeed well down into the green before you reach it. If you think that there might be some pretty good bumps down there, be at or below maneuvering speed *before* you encounter them.

The winter winds bring cold weather with them, and that's the time to switch to winter-weight oil, check the heater, restudy cold-weather starting procedures, pledge anew to get all ice, snow, and frost off the airplane before flying, and pay special attention to places (such as prop spinners) in which rain could collect and then freeze solid.

One other thing: if you live north or west of Florida, you can sometimes get a great winter tailwind to the Sunshine State. Better plan on staying a few days if you want a tan, though, for any winter wind that blows you to Florida in a great hurry also takes large volumes of cold air with it.

30.

... AND CROSSWINDS ...

Somewhere north of your northernmost sectional chart an ill wind waits to blow little to no good as it sweeps inevitably toward you. Wherever that crosswind lurks, scores of pilots are even now flight-planning themselves toward a confrontation with it that is keenly anticipated by some, awaited with terror by others, and totally unexpected by a few. There is no getting around it, however: every pilot must do occasional battle with a rough-and-tumble wind that blows askew to the runway − yet it is an engagement that should be treated as the inspiring challenge that it certainly can be. The ability to steer an airplane through acceptable landings and uneventful rollouts regardless of wind-to-runway alignment is a technique that can be learned, practiced, and exhibited with vast pride.

The problems are simple enough: an airplane acts like the part of the air in which it flies, and, like Mary's little lamb, anywhere the air goes, the airplane is sure to follow. To shift the metaphor a bit, an airplane is like a goldfish swimming through a bowl that is being carried by a running man. The speed of the goldfish through the water represents true airspeed, while groundspeed and track are influenced by the speed and direction in which the man (or the wind) is moving. All of this would be largely academic − except for the frequent necessity to aim an airplane elsewhere in order to have it arrive at a distant point − if it were not for the very real requirement that airplanes be pointed very near their direction of travel at the moment that their wheels touch the runway. Even the toughest of wheeled vehicles are meant only to go forward or back, not sideways. Airplanes are in an even more delicate situation because of the inherent instability of three-wheeled vehicles. Airplanes have wheels simply so that they can be driven to the end of the runway and then rolled along it to gather sufficient speed for flight. The effectiveness of an airplane as a ground vehicle is akin to that of a beginner on skis: as long as each sticks very close to the original plan, all should go well enough. Plans may come to a sudden change, however, if the tower says, "Cleared to land runway 27; wind is 340 degrees at 20 knots." Unless the pilot can somehow pluck his aerial goldfish from its sideways slide as the wheels touch, he will be asking for the impossible: an airframe with enough qualities to protect the pilot from himself.

There are three basic methods employed by airmen in their pursuit of zero drift at touchdown − the *crab*, the *slip*, and a *combination* of the two. The crab simply uses the old navigational device of aiming over here when you intend to go over there. The pilot flies downwind, base, and final with the airplane aimed more or less to windward of his desired course, and just before touching earth he kicks out whatever crab-angle correction he was holding by stomping on the downwind rudder and adding just enough opposite aileron to keep the wings level. He finds himself inches off the ground with an airplane that is pointed precisely in the direction in which it's traveling. Before congratulating himself prematurely, however, he had better remove those last few inches between tires and runway, or else the blowing crosswind will override the airplane's momentum and the sideways drift will begin again. The kicking-out-the-crab method is really hellaciously difficult because of variable winds and turbulence and the need for superhuman timing of flare height and rudder push. Very few people try it even once, and those who do attempt it seldom come back for more.

If the crab method appears reasonable on a blackboard, however, then the second method — the slip — must at first look outrageously silly. The principle here involves utilizing the lift force produced by the wing to counteract the force that drifts the airplane sideways. If the force being generated by the wind can be canceled out by varying the direction of the usually vertical lift component (a fancy way of asking the pilot to roll the airplane into a bank), then the airplane will stop moving sideways. For a graphic illustration make like an airplane with your hand and place a pencil vertically between the two middle fingers to represent the force and vector of the lift generated by the wing. Now move your hand to a spot 1" or so from the wall and hold it there: the space between your hand and the wall represents the crosswind component. By rolling your hand slowly into the "wind" (and thereby tilting the pencil toward the wall) a position will be reached at which the pencil's tip touches the wall — and if your hand were really an airplane, this would be the moment of balanced forces and zero drift.

Watching a hand go through this exercise a few times brings to mind several observations. First is the necessity to somehow stop the airplane from turning, which is what it will do whenever a wing is lowered. The solution is to apply sufficient opposite rudder to prevent the airplane's nose from moving — an application that results in crossed controls and the slip from which the exercise gets its name. If you move your hand somewhat farther from the wall, you'll notice another problem: there comes a point at which, regardless of the pencil's bank angle, its reach is insufficient to bridge the distance (counterbalance the force) between hand and wall. This demonstrates that there is a theoretical limitation in controlling drift with tilted lift vector by showing the point beyond which even a 90-degree bank can no longer throw enough energy outward to balance the wind.

In actuality, however, the crosswind limitations of the slip method are determined by two other factors. The physical properties of the airframe are often the first. Maximum bank angle simply cannot exceed the angular relationship between the upwind tire and its wingtip, or the pilot will drag the wing (or tip tank or engine pod) before he gets the wheel on the ground. A high-wing airplane can obviously do more rolling than one with a low wing or other obstructing appendages. The second crosswind limitation involves the rudder and its ability to hold the airplane's nose straight despite the lowered wing. Maximum crosswind factor has been met when the opposite rudder hits the stop. Most single-engine general-aviation airplanes will run out of sufficient downwind rudder before meeting the limits of their airframe's angular abilities.

Two-engine airplanes (except for the centerline-thrust Cessna Skymaster) are more versatile in a crosswind, because the pilot can utilize asymmetric power to further enhance his rudder's ability to stop the turn. A pilot in a twin who uses power on the upwind engine in addition to rudder displacement can handle much more drift than his single-engine counterpart, and in almost all instances his crosswind limitations revolve around the physical ability to get the wing down. The biggest crosswinds can be handled by a high-wing, wide-gear twin, while only lighter crosswinds can be attempted in weak-rudder, low-wing singles. Even though there are limitations to the slip technique, it is almost always superior to the "kick out the crab" exercise, because it allows the pilot a clear gauge of his timing and coordination and eliminates his dependency upon one perfectly planned push on the rudder. In spite of its advantages, however, pure slip-type crosswind landings are seldom performed.

Although it may be a matter of semantics, the problematic definition of exactly where a crosswind landing begins (on downwind leg, base-to-final, half a mile out?) eliminates the pure slip from serious contention. A pilot could actually use a lowered wing during en-route flight. in order to counteract wind-induced sideways movement, but the silliness of flying 400 miles with one wing down is obvious. Crossed controls provide an aerodynamically inefficient, uncomfortable ride that should be avoided whenever practical. Besides discarding some lift, inducing additional drag, and causing all aboard to lean left, airfoil disturbances and erroneous airspeed indications are a few other reasons to keep the slip portion as brief as possible — which is why most airmen utilize the combination crab/slip technique in crosswinds.

Crabbing the airplane slightly to windward on downwind, base, and a portion of final approach will handle the pattern chores involved in tracking a desired course in spite of quartering gusts. Some pilots recommend transitioning to the slip method quite a way out on final in order to get the airplane's attitude firmly established, but this advice should be tempered by one overriding concern: the wind encountered two miles out at 1,000' is quite different from the one the pilot must contend with as he crosses the numbers. Any established bank angle and rudder deflection will certainly have to be altered as the descent continues. The more experienced the pilot, the longer he can continue his crabbing final approach before going wing down to a driftless landing. Less-experienced people *should* go to the lowered wing somewhat sooner (it gives them more time to practice aileron/rudder cross control), but the discomfort that sideward-sliding crossed controls can produce in a cabin full of passengers must be remembered.

A discussion of airspeed selection and use of flaps or no flaps will certainly bring pilots piling out of the hangars, but this much is certain: the airplane should be flown primarily by reference to airplane attitude and engine power, since airspeed indications are usually somewhat inaccurate. Although the slipping maneuver does increase stall speed, the amount of increase can generally be disregarded, since extra airspeed added to compensate for the almost invariable gustiness of a crosswind will be a more than adequate cushion. Touchdowns are best accomplished with an attitude-type drive-on-the-upwind-wheel landing with sufficient airspeed to maintain control and allow the pilot to pick the exact driftless moment to plant the wheels. Although such landings require somewhat longer runways (not too much more, though, because all that wind has diminished groundspeed), the additional use of concrete is made worthwhile by the fact that the pilot has a firm, flying airplane as he fiddles with aileron and opposite rudder.

The selection of full flaps usually complicates things by producing lots of nose-down pitch during final approach, a condition that must be eliminated during flare. The pilots of many larger transport aircraft, in fact, secretly use lesser flap settings (the technique is not FAA-approved, because there are often no official performance charts printed for landing with less than full extension) to provide a nose-up descent if they encounter a particularly violent wind, eliminating the wrestling required to pick up the nosewheel — a bout that would occur at the worst possible instant, that at which the pilot needs to set the machine on the runway at precisely the properly timed moment. In light of the groundspeed reduction caused by a heavy wind it is difficult to justify the increase in flare and timing problems that accompany full flaps. It seems well worth a few knots more airspeed to have an airplane whose pitch and attitude and controllability are matched to the type of touchdown needed.

The amount of bank angle to use is a delicate question that can best be answered by continuously putting down more and more wing until the runway stops its apparent move upwind. Simultaneous application of opposite rudder is needed to keep the nose parallel with the runway: a nose pointing toward the downwind side shows too much opposite rudder; pointing upwind shows too little. If the fuselage is aligned with runway but the airplane is drifting upwind, there is too much bank angle. A drift to the downwind side signals the need for just a bit more aileron into the wind.

The landing itself is accomplished by first placing the upwind wheel on the ground after it has been properly aligned by rudder input and made driftless by virtue of the airplane's windward bank. The task of putting the airplane on the other two wheels can be touchy, especially since it is necessary to center the rudder pedals before putting the nosewheel on the ground if your aircraft has its groundsteering mechanism linked firmly to its front strut. The manufacturers who have tied rudder to nose strut make crosswind touchdowns and rollouts markedly more difficult.

There have been many dramatic demonstrations of the fact that crosswind landings aren't over until the airplane reaches the hangar. Once your machine is firmly on the ground, its sideward tendencies will be countered by the friction of the tires against the runway, although you should help with continued application of aileron into the wind and (if possible without simultaneously skewing the nose tire) opposite rudder. The strongest crosswinds always seem to blow just as the front moves through or the blizzard strikes or the squall line brushes past, and the more slippery the runway surface, the more difficult it is to keep the aircraft's track and fuselage fore-aft axis steadfastly aligned. If braking action is truly nil, the job goes beyond Herculean and becomes quite impossible: the only way to keep an airplane on a crosswind runway when tire-friction coefficient is zilch is by reverting to the crab method of drift control and pointing the airplane somewhat upwind — a maneuver that results in a definite skid, stress, strain, clenched-teeth risk, and unquestionable pilot fatigue.

There are as many ways to recover if things begin to go wrong on rollout as there are ways to do the wrong thing at the right time. Techniques range from full-power go-arounds to semicontrolled ground loops, and all of them have a way of working sometimes and not others. The surest recovery method is to keep everything perfectly in line all through the landing and rollout, but, barring this, the most appropriate rule seems to be to avoid any huge control or power inputs or changes during attempts to salvage a crosswind touchdown. Stomping on the brakes is almost always doomed to failure, while a late go-around might be more dangerous than drifting off the runway.

Of all the day-to-day experiences that airmen have, handling the crosswind landing can be the most enlightening maneuver that he is called upon to execute. Regardless of technique or time or type, a crosswind is the fastest humbler of pilots around — a charm well in keeping with its ability to make a nicely done landing the reason for walking around the ramp with a satisfied smile. The advantage seems to lie with those men who have learned how to utter a few magic words, the pronouncement of which appears to separate the masters from the misfits: "Hey, tower," they say as they roll out smoothly, "I'd like to taxi back to try a few more."

31.

... AND THOSE TERRIBLE DOWNWIND TURNS

While the roundness of the earth is now taken for granted by almost everyone, the similarity of downwind and upwind turns for some reason still remains a point of argument. Whole generations of disputants on both sides of the question have failed to lay it to rest, and it would be presumptuous to claim to do so here when greater men than we have failed on this battlefield. Our word, therefore, may be taken by those who obstinately differentiate between downwind and upwind turns as a mere statement of policy by *Flying* magazine.

The argument that downwind turns are dangerous proceeds from the incorrect assumption that kinetic energy is always to be measured with respect to the ground. If this is the case, then an airplane traveling downwind clearly has much more kinetic energy than one traveling upwind, because it is moving faster with respect to the ground. Now, the argument goes on, the airplane disposes of two types of energy: kinetic energy, which is the energy of a mass in motion, and potential energy, which is the energy of a mass raised above the ground. If you are flying upwind, you have little kinetic energy, and so-and-so much potential energy, depending on your altitude (the higher you are, the more potential energy you have). Suppose that you make a 180-degree turn so that you are now flying downwind. Your groundspeed will have increased considerably, but if your altitude has remained the same, where did you get all this additional kinetic energy? From your engine, which, as you turned downwind and lost indicated speed, caused the airplane to accelerate and maintain its original speed — the trimmed speed, say. The engine can only accelerate you at a certain rate and no faster, because it produces only a limited amount of power. If you turned quickly enough, therefore, the engine would not have time to bring you up to trim speed as you turned downwind, so you would note a loss of speed, or, if you held speed constant, a loss of altitude (because you traded some potential energy for kinetic energy). If you happened to be flying just above the stall, you might stall; if you happened to be flying just above the ground, you might crash. Does this argument sound pretty good to you so far? If it does, then you misunderstand the most basic of aeronautical relationships — that of the moving airplane to the air. That you should misunderstand it is not surprising, however, considering how poorly it is taught and how poorly it is understood even by many instructors.

The first axiom of airplane behavior is that it is unaffected by a steady wind, whatever the wind's velocity. A steady 100-knot gale is equal to a dead calm. The difference is significant only when you touch the ground — obviously, if you are going to land in a 100-knot gale, you had better do it upwind. As far as maneuvering in the air is concerned, at any altitude wind speed is meaningless. The reason is simply that the energy of the airplane is measured with respect to the air, not to the ground. An airplane flying upwind and another flying downwind at the same IAS are in identical states of flight, and they may turn equally sharply in any direction they please, and neither pilot will be able to guess from the behavior of his airplane which way the wind is blowing.

The situation is comparable to that of a man running along the deck of an ocean liner. The downwind-dangerous argument proceeds on the assumption that the kinetic energy of the running man is measured with respect to the sea, not to the ship. Suppose that he is running at 10 mph and the ship is moving at 10 mph; suppose that he is running in the same direction that the ship is going and suddenly turns about and runs the other way; suppose that he collides with a stanchion. If you were this unfortunate gentleman, would you prefer to have hit the stanchion while running toward the bow or toward the stern? Those who measure kinetic energy with respect to the sea must choose to hit the stanchion while heading aft, since they would then have no kinetic energy and could accordingly feel no pain. Consider another demonstration of the fallacy: an airplane is flying into a strong wind at zero groundspeed and at a certain altitude. The pilot pulls up: he gains altitude and loses speed. He now has *more* potential energy, but where did he get it, since he had *no* kinetic energy to start with? Does he now have *negative* kinetic energy? If so, what is negative kinetic energy?

To return to reality, by measuring the airplane's kinetic energy relative to the air rather than to the ground, you find that that theory corresponds exactly with your own experience of flight. Even with the engine stopped an airplane loses no more altitude in a downwind turn than in an upwind one. Anyone who cares to do so may wait for a good windy day with 30 or 40 knots at 6,000 feet, and go up and try it for himself. Make a series of 360s and see if the airplane loses altitude as the wind comes around behind it. To be particularly scientific, reduce power to the minimum necessary for level flight. You in effect eliminate the accelerating effect of the engine, since there is no extra power left for acceleration even if the airplane does slow down in the turn. A little extra power may be needed through the entire turn because of the added drag of the G-load, but the additional power will be constant throughout the turn and will not depend upon the direction of the wind. It can be readily demonstrated for the mathematically minded that the forces acting upon a turning airplane in a steady wind are independent of the wind direction. (A detailed demonstration may be found in the June 1939 issue of *Aero Digest*.)

Once it has been understood that the airplane flies *in* and with *respect* to the air and altogether without respect to the ground, the stage is set for another fallacy. This is the Great Dying-Wind Conundrum, and it, too, like the poor, is always with us.

The fact that the airplane flies solely with respect to the air leads some to suppose that it may be thought of as *a part of the air mass*. They imagine, therefore, that an airplane with zero groundspeed flying into a strong wind will continue flying if the wind suddenly dies to nothing. In fact it will not: it will fall − more or less whip-stall into a dive − and then recover, time and space permitting. Luckily this does not befall us very often in nature.

The fact that the airplane's flight is referable to the air does not mean that the airplane is part of the air. It is not: it is a separate mass that is quite a bit denser than the air. You probably noticed that throughout the preceding discussion steady winds were carefully specified. Unless a wind varies its velocity so gradually that engine power masks changes in airspeed, changes in wind velocity will register on the air speed indicator. Because the airplane is more dense than air and is so shaped as to minimize the effects of air upon it, it is slow to respond to changes in the velocity of the relative wind. Returning to that ocean liner: if it happens to run aground, the passengers will tumble down, since they are not rigidly attached to the ship, and will keep moving forward even when the ship has suddenly been stopped. The airplane is not rigidly attached to the air, and, if the wind velocity suddenly changes, the airplane tends to maintain its *groundspeed* − not because the ground has any effect on the airplane, but because it, like the airplane, is not rigidly attached to the air.

Normally winds do not suddenly die; they may fluctuate rapidly, however, which is the condition referred to as "gusty air" and that we dealt with by adding a fudge factor to our approach speed. The point of adding a 10-knot safety factor to the already existing difference of, say, 20 knots between approach speed and stalling speed is to make sure that a wind-speed drop that could cause you to stall is highly improbable. A 30-knot drop in the wind is much less probable than a 20-knot drop, and the function of the fudge factor is merely to *improve your chances* of not stalling.

There is one thing that *can* make downwind turns more dangerous than upwind ones. Suppose that you are flying an airplane that stalls at 50 knots; you are doing 70 knots IAS into a headwind reported as 10 knots gusting to 25. Even if the wind died out entirely, you would maintain flying speed. But suppose that during a period of relatively steady 10-knot winds you turned downwind, still maintaining 70 knots. If a 25-knot gust now struck you from behind, you would stall at least momentarily. This is an unlikely situation, but it is a nice example because it contains opportunities to fall into both fallacies − the downwind-turn fallacy and the dying-wind fallacy. If you can imagine your way through this one and come up with the right result, you will understand the relation of wind to airplane better than a lot of people. You will probably also understand why the fudge factor can be even more important if you're flying downwind in gusty air than if you're flying upwind.

If the downwind-turn fallacy really is a fallacy, then why do so many high-time pilots, especially ag pilots, swear up and down that their planes perform worse in turning downwind than up and why do they tell you about the startling regularity with which they buried victims of downwind turns back in the old days?

Several reasons: for one thing people believe what they want to believe and even manage to find evidence for things that aren't true. You don't have to go far to find evidence for the dangers of low-altitude, low-speed downwind turns. Any low-speed, low-altitude turn is dangerous; a turn downwind at low altitude, however, reduces the climb angle of the airplane, giving the impression of a loss of performance, and could lead to collision with some obstacle. Many a downwind turn has followed an engine failure on takeoff and ended in disaster not because it was a downwind turn but because it was a no-power turn executed at low altitude with insufficient airspeed. Contact with an obstacle or the ground after a downwind turn takes place at a higher speed than if one were traveling upwind and is therefore more dangerous.

The most important danger in climbing downwind turns, which are the kind that ag pilots frequently make in order to clear obstacles at the end of a field, is the wind gradient. Furthermore, ag pilots often work in the morning and the evening, when the wind is calm near the surface. A little higher up − at 300 feet, say − the wind velocity is commonly more than twice what it is at 20 feet. If an airplane is climbing into a wind that increases with altitude, it gets a boost in climb angle, and, if the wind increases rapidly enough, it may even get a temporary boost in indicated airspeed. Climbing downwind, the opposite happens: the boost becomes a sag, and a strong wind gradient can cause a loss of indicated airspeed. Add to this the high power required by a climbing turn, the rise in stalling speed associated with banked attitudes, and the weird visual impressions received by the pilot in executing a downwind turn at low altitude and you have the reason why the subject of downwind turns has given off so much smoke for so many years. Where there's that much smoke, there has to be some fire − and there is. The old rule of shunning low-altitude downwind turns and most particularly climbing downwind turns is a good one: only the explanation usually given for it is bad.

32.

WHEN THE SUN GOES DOWN

Night flying is sufficiently similar to daytime flying as to be sometimes dismissed as "the same thing." Many beginning pilots have been surprised to find on their night checkout flight that, apart from a few adjustments in their landing-approach procedures, there was little difference between after-dark aviating and everything that they had learned to do in daylight. On the other hand, night flying is treated in some countries as a first cousin to IFR flight: in Mexico, for instance, all night flight must be done under IFR. Evidently opinions vary as to the difference a day makes.

In certain respects night flight is easier and pleasanter than daytime flight. Traffic is usually easier to spot at night; the air is usually smoother and cooler so that the airplane is more comfortable and performs slightly better; radio traffic is sparse; traffic patterns are nearly empty. There is no glare. Instrument scanning is easier — if cockpit illumination is good — and visibility is often better than in daytime, both because airborne dust is settling and because distant towns and highways that fade into the haze in daytime stand out clearly at night as lights against a black background.

On the other hand, certain difficulties are peculiar to night flight. Judgment of the landing approach and roundout point is harder. Fatigue and the droning of the engine conspire to undermine one's alertness, and there is little of interest in the visual field to hold attention. In marginal weather, when VFR flight is still possible in daylight, cloud avoidance may be difficult at night unless there is a bright moon, and IFR may be the best way to go. (At night IFR delays and detours are fewer than in the busy daytime.)

What makes some countries view night flight as a special genre requiring special skills is, of course, the fact that forced landings and precautionary landings that one might bring off safely in daytime are nearly impossible to do at night. Many pilots in the United States, where no distinction is made between day and night VFR in the rulebooks, nevertheless refuse to fly single-engine at night. Others are willing to take the chance. Debate is endless and pointless about just how much of a chance it really is — endless because statistics are so vague and ambiguous and pointless both because statistics for the whole fleet have no bearing on a specific aircraft and because what really makes people's minds up is not numbers on a piece of paper but personal qualities such as self-confidence (considered recklessness by those who lack it) or indifference to danger. Some pilots have never had any engine trouble at all, though they have flown thousands of hours, and their cars have never broken down completely either, so they press on when the sun goes down. Others have had bad luck with machines, don't trust them, and hedge all their bets. The ultimate decision lies with the pilots themselves, and the same rule applies here as everywhere: if you don't feel comfortable doing it, don't do it just because somebody else does. If you feel comfortable and your passengers also do, go ahead.

If you go, there are a number of precautions that it would be foolish not to take. One is to flight-plan a more generous fuel reserve than you ordinarily do. There are many reasons for this, the most fundamental of which is that, since darkness robs you of the option of making precautionary landings, makes pilotage much more difficult, and puts a heavier load on the electrical system (a failure of which could leave you without navaids), you should give yourself a little more backup than you usually need. Another reason is that fuel is available at fewer airports at night, and one is tempted, once airborne, to go as far as one can. Temptation goes farther than fuel.

Another indispensable piece of self-defense is a flashlight. You need it for the preflight — be sure to verify the security of gas and oil caps, particularly, and removal of the pitot cover — and it may come in handy in the cockpit for map reading or finding things on the floor. There is also the off-chance that your instrument lights may go out; if you land at an unattended field, it would help you find tiedowns or chocks. Checking the fuel tanks with a finger is fair enough, no doubt, but you need a light to check fuel color and also to check the runoff from the drains — assuming that you use a plastic cup — for contaminants. Finally, a flashlight will help you get the key into the door lock.

On takeoff be prepared to rely on the instruments if the runway aims out over water or empty fields. Instrument capability is not indispensable for night VFR, but it can help a lot.

Some people like to keep the panel lights quite dim on the grounds that they can see lights outside the airplane better that way. The original reason for the use of red panel lights is that red light interferes least with the eye's adaptation to night vision, and, whatever may be the "official" explanation of the present switch to gray panels and white light, it may well have to do with a general shift toward increased reliance on instruments and avionics and a lessening emphasis on external reference in night flying. When the weather is CAVU, a dim panel is still preferable: only in instrument flight need one turn up the panel lights. Keeping the lights dim can be rationalized on the grounds that, if a forced landing were necessary, the best possible night vision would help.

En route, it is advisable — again in order to make up for missing margins of safety — to choose high cruising altitudes. The reasons are several: range is greater at higher altitudes; gliding distance is greater; pilotage and radio navigation may be easier; and it is easier to stay within gliding distance of a lighted airport most of the time. In still, cool night air, climb performance is good, and going to a higher altitude usually does not involve a great loss of time or fuel.

Some pilots will use a different route at night than in daytime — one that keeps them within reach of an airport as much of the time as possible. If you trace a zigzag route on a chart and compare its length with a direct one, you find that zigzagging adds surprisingly little mileage: for instance, a course comprising a series of 25-degree zigzags is only 10% longer than a direct one. Most light airplanes will give a glide ratio of eight to one or better: if you fly at 10,000 feet agl, and manage to stay within 16 miles of some airport all the time, you may as well be flying in daytime. It is not always possible to stay within 16 miles of an airport − especially in the western part of the country − but by planning a route that threads its way along a string of lighted airports or roads you can reduce the amount of time during which you are out of gliding range to a minimum.

Highways look like tempting emergency-landing strips at night and following them may simplify navigation, but, unless the highway is wide enough to land on without hitting the power lines that are likely to run unseen alongside it, it may do little to decrease the hazards of a forced landing at night. With practice you get quite good at picking up the alternating green and white beacons of airports − or green and double white of military ones. You also learn by puzzling experience that if you think that you see something and stare at the object, it disappears, but if you look slightly to one side of it, it becomes visible again. The reason is that the elements in the eye best suited for seeing at night are found in the peripheral parts of the retina, while the central part of the retina − the part that you are using when you stare directly at something − gives superior definition and color perception in bright light but is inefficient in darkness. If you are hunting for a rotating beacon and think that you glimpsed it, look a little to the side of the place where you think it was, and you may see it again.

It is important at night to keep closer track of one's course on the charts than one might do in daytime: in particular terrain height and the locations of obstacles are good to know. Use a white light for reading charts: much of the information is in red and fades under a red light. Familiarize yourself with the data for your destination airport well in advance; of particular note is altitude. Get an altimeter setting before landing. When entering the traffic pattern, use a standard altitude − 800 feet, as a rule − and fly it with precision.

Most pilots use the landing light for landing − naturally. Most instructors, on the other hand, suggest that you practice landing without lights. The portion of the runway that is illuminated by the landing light seems higher than the black hole surrounding it, and this can lead to unduly hard landings. To focus one's attention on the area immediately in front of the airplane is poor practice on any landing, but this is exactly what the landing light encourages. If you practice landing without the light, by looking at the distant end of the runway and using the rising light lines on either side as peripheral cues for the flare and touchdown, you will find that you flare at more nearly the correct height and that your landing attitude is better than if you try to put the plane into the pool of light created by the landing-light beam.

Besides using the runway lights rather than the airplane's light as landing references you can avoid trouble by flying a carefully standardized approach pattern, using your altimeter to verify your rate of descent. If you fly the downwind at 800 feet, plan a square descent beginning at a point opposite the touchdown point. When the touchdown point is off your wing, reduce power for a descent of about 500 fpm; when the touchdown point is about 40 degrees behind you, turn base; you should be at around 600 feet agl. Rolling out on final, you should be at around 200 or 250. Fly the pattern close enough to the runway to make possible a square approach with a constant sink rate; keep the sink rate steady and keep your airspeed right on the correct value − don't let yourself compensate for whatever uncertainties a night landing may entail by adding a little margin to the approach speed. The result is to prolong the flare and increase the likelihood of misjudgment.

In theory, a landing light should throw a narrow, bright beam, and a taxi light should be broader and more diffused. In most airplanes, however, taxiing on a poorly lit, unfamiliar field is misery. You simply have to be cautious and proceed slowly. Many smaller strips have a few runway lights out and no taxiway markers: if you don't look carefully where you're going, you can end up in a ditch or stuck in soft ground.

If you land at an unattended field, you may be tempted to leave the brake on and hope that the wind doesn't come up during the night. Brakes, however, do not always stay on: their hydraulic pressure may bleed down in an hour or two. It is important, therefore — even though the job sometimes seems nearly impossible — to find a tiedown or at least two secure chocks. It may involve shutting down temporarily, searching on foot for a tiedown, and taxiing over to it afterward. The inconvenience is painful if you're tired and longing for a bed, but it's worth it for your own sake and others'.

One final caution: night is the time when there is the greatest likelihood that one will have set out with a couple of drinks under one's belt. It is so obvious that this is not a good idea, yet every year, in airplanes as in autos, alcohol is implicated in a disproportionate number of accidents and deaths. Night flying is perhaps a hair's breadth more hazardous than daytime flying; with the aid of a few drinks, however, night wins the danger sweepstakes hands down.

33.

INDUCTION ICING

We used to talk about "carburetor icing," but with all the modern fuel-metering devices used today, a more up-to-date phrase describes this flight condition as "induction-system icing." This term includes not only carburetors but fuel injection and also covers all the vulnerable parts of the induction system in which ice can accumulate — the air filter, bends in the system, and the critical areas of the fuel-metering accessories themselves. If pilots understand what happens when the fuel-metering device (carburetor or fuel injector) injects fuel into the air being sucked into the induction system by the engine as it operates, they can take suitable precautions to eliminate or — preferably — avoid induction-system ice altogether.

Under certain moist (and "moist" is a key word here) atmospheric conditions, with air temperatures ranging anywhere from 20 to 90 F it is possible for ice to form in the induction system. The rapid cooling in an induction system using a float-type carburetor is caused by the absorption of heat from the air during vaporization of the fuel and is also due in part to the high air velocity, which creates a low-pressure area through the carburetor venturi. As a result of the latter two influences the temperature in the mixing chamber may drop as much as 70 F below the temperature of the incoming air. If this air contains a large amount of moisture, the cooling process can cause precipitation in the form of ice, generally in the vicinity of the butterfly, which may build up to such an extent that a drop in power or even complete engine stoppage could result. Indications of icing are a loss of rpm with a fixed-pitch propeller and a loss of manifold pressure with a constant-speed propeller, with an accompanying drop in airspeed for both types.

The thinking pilot will anticipate possible icing and utilize heat *before* the ice forms. However, should he fall to anticipate icing in an aircraft with a float-type carburetor, he must definitely use the full heat position once the ice has begun to form in order to be sure of eliminating it. Using full heat will initially cause a loss of power and possible engine roughness; heated air directed into the induction system melts the ice, which goes into the engine as water, causing some of the roughness and even more power loss. Unless the pilot knows what is actually happening, the stress and confusion of the situation may tend to frighten him away from using heat and could result in the loss of the engine to icing. In using heat some power loss is incurred by going from ram-air position to the less direct flow of the heat position. Carburetor heat or alternate air also creates a richer mixture, which may cause the engine to run roughly — particularly at full heat. The mixture should be adjusted lean (assuming that you're running at cruise, though at lower power conditions − in the traffic pattern, say − it may be impractical to lean the mixture). Do not use heat during takeoff or climb, as it may bring on detonation and possible engine damage. An exception to the latter might be in the severe temperatures of the Arctic, which call for special knowledge.

If you are wondering how long to continue the use of heat, it depends on the severity of the icing condition. If icing is severe, heat should be used as long as flight continues. Despite the temporary roughness and moderate power loss no amount of heat can damage an engine running at a cruise setting of 75% power or less. Lycoming has found in its flight tests on various models of engines with special detonation pickups that at *cruise power* with full heat or alternate air they have never experienced detonation or damage to an engine. If this is difficult to believe, remember that a turbocharger heats the induction air considerably hotter than carburetor heat or alternate air without causing any detonation or damage to the engine. And, after all, the principal concern of a pilot under icing conditions should not be the possibility of detonation from the heat at cruise power but rather to keep the engine running, no matter how much heat is required.

If the airplane does have an induction-air or carburetor-air temperature gauge, the thinking pilot who *anticipates* the possibility of induction ice can prevent it by maintaining a minimum of 90 F during cruise and letdown. Flight tests with good instrumentation have generally verified that, with either fuel injection or carburetion, maintaining a minimum of 90 F hot-air temperature in the induction system *before* induct icing conditions occur would prevent the formation of induction-system ice *as long as it was anticipated by the pilot*. If the pilot failed to anticipate icing, however, an undetermined amount of heat may be required to eliminate it once it has accumulated. Any aircraft without an induction-air gauge must use either the full heat or full cold position, as an unknown amount of partial heat can actually *cause* induction ice in the float-type carburetor, particularly if moisture is present in crystal form in the incoming air that would ordinarily pass through the induction system without any problem. Partial heat melts these crystals, and they form carburetor ice in the venturi. At a temperature of 14 F or below any moisture in the air is frozen.

A recent aviation-magazine article has claimed that all pilots are taught to use carburetor heat during the landing configuration and must be sure to get rid of it in case of a go-around. True enough − but the author's explanation of the need to get rid of the heat on a go-around was: "The real danger is that the application of carburetor heat increases the richness of the mixture, and upon sudden application of full throttle for a go-around it would be possible to flood the engine, causing complete engine failure." This is incorrect. Although heat does make a richer mixture, it does not "flood the engine." The basic reasons for removing heat on a go-around are: (1) Loss of power with full heat on becomes critical at low altitude and low airspeed. (2) There is a danger of detonation and/or engine damage if full heat is used during a go-around at takeoff or climb Power on a high-performance powerplant.

Lycoming has investigated forced landings caused by carburetor ice in single-engine aircraft. Several of these involved solo students flying the same route and altitude as their flight instructor, who had a second student in the area at the same time. In some cases investigation revealed that the instructor was flying without carburetor heat but had at least leaned his mixture and *did not* experience any carburetor icing. On the other hand, the student was flying at full-rich mixture and without carburetor heat. The refrigeration action from full-rich mixture in marginal conditions created enough difference in induction-air temperature to bring on icing.

The float-type carburetor can incur ice in VFR flight conditions with the right combination of moisture, temperature, and fuel mixture without heat. In the case of fuel-injection and pressure carburetors it is IFR conditions that generally cause induction-system icing. Icing will not choke a fuel-injected engine at the venturi, but in extreme cases slush snow can cause blockage of the induction system despite the alternate-air system. Engine manufacturers have also run into special induction-icing problems with fuel injection in light, powdery snow. One such case that Lycoming investigated involved a twin- engine aircraft with fuel injection that flew into light snow and incurred power loss in both engines due to induction-system ice. Along with his application of hot air the pilot went to full-rich mixture and increased his power. The rich mixture and increased power actually made the power loss worse under these conditions, because the increase in fuel vaporization actually *intensified* the refrigeration action at the injector nozzle. Engineers later duplicated this icing condition in flight and were able to avoid the dangerous power loss by leaning the engines close to peak exhaust-gas temperature and holding the established cruise power. Leaning should be accomplished in each case *after* application of carburetor heat or alternate air.

The typical turbocharger powerplant should have no problems with induction-system icing except in extreme conditions, because of the high temperature of the induction air when the compressor is running. Slush snow, however, can block the air filter if alternate air is not readily available. The pressure carburetor is similar to the fuel injector in that it is not very vulnerable to icing other than that brought about by filter blockage. In a float carburetor, the fuel jet is ahead of or just below the venturi and throttle butterfly, which means that fuel is squirted directly into the worst possible place for icing − the carburetor venturi. On the other hand, a pressure carburetor's jets are squirting fuel farther downstream − beyond the venturi refrigeration chamber − which accounts for the decreased likelihood of icing in this type of system.

Remember, however, that most pressure carburetors have automatic mixture controls. Any application of heat on the ground will affect the automatic mixture control unit, making it temporarily unpredictable in its effect on the carburetor in flight. If you check the heat system during a run-up, you must wait at least two minutes before takeoff in order to avoid an erratic fuel flow because of the effect of the heat on the automatic mixture-control unit.

34.

MULTIENGINE EMERGENCIES

The two most dangerous aspects of twin-engine operations are airplanes that can't fly on one engine and pilots who can't fly on one engine. Assuming that a twin is healthy enough to stay up with one feathered and that you haven't overloaded it so grossly as to compromise this ability, the handling of mechanical failures is the most basic skill that a multiengine pilot can develop. In fact, it is almost the only skill that he can develop beyond those that he has assumedly learned and polished on single-engine airplanes. To be honest about it, the major differences between a light twin and a sophisticated single are that the potential for disaster is far greater in a twin and that, if disaster threatens, it requires the twin pilot's immediate and precise attention.

Sound frightening? It shouldn't, for there is a foolproof way to handle light-twin emergencies that can be practiced constantly − every day on every flight − and that is so easy to assimilate and remember that after awhile you'll wonder why you ever worried about the specter of an engine failure on takeoff.

In an emergency of any kind the dangerous extremes are sheer inactivity and frantic, misguided haste. Let a fire light glow or an engine lurch its valve train, and many pilots will sit as stunned as a poleaxed bull, watching the oil pressure dribble to zero and the airspeed decay to stall; they are suddenly unable to remember their own names, never mind the feather checklist. That's bad. Also bad is the theory, "When in doubt, run in circles, scream, and shout": adherents to this modus operandi tend to feather the good engine or to pull the firewall shutoff when all that they needed to do was to switch tanks or to retract the gear when they meant to bring up the flaps.

The perfect mean between these extremes is the all-purpose, never-fail, every-ready, easy-to-remember emergency procedure − a single solution for all light-twin mechanical failures in all flight regimes. The trouble has been, however, that many instructors and examiners have advocated different procedures for different airplanes and even different procedures for different situations in the *same* airplane.

The next time that you feel hearty enough to practice multiengine emergencies, get yourself a good instructor or a highly competent safety pilot and tell him to pull back engines unexpectedly during takeoffs, climbs, cruises, and approaches and to hit you with a few single-engine waveoffs as well. (It goes not quite without saying that this means pull *back* an engine, not pull: zero thrust is adequate, thank you, and there is little excuse for feathering an engine at low altitudes in training. After all, you don't light a bundle of oily rags in the nacelle when you practice inflight fire procedures, do you?) When he sneaks that throttle back, hit him with this six-step procedure, and never again will that sickening yaw strike terror into your twin-engine heart: *power, flaps, climb, gear, identify, feather.* Power: The pedestal levers all go full forward − mixtures, props, and throttles. Flaps: The flaps come up to normal takeoff setting, and, if they're already there or fully up, you simply confirm that fact. Climb: The airplane is rotated to a climb attitude, the vertical speed indicator is checked to ensure that any descent trend is reversed; and the ASI is checked to confirm that you've nailed your best-single-engine-climb speed. Gear: Retract gear. Identify and feather: Confirm which engine has gone dead by the hoary old "dead foot, dead engine" method, then sing out your feathering procedure to the check pilot. Finally, clean up the flaps once climb is established.

Easy? In its wordiest form − the one you should recite out loud as you run through it − the procedure goes: "Power up, flaps to takeoff, climb attitude, gear up, identify, feather."

This appealing little formula actually has three lovely faces. It is simple − uncomplicated by exceptions or yes, buts, applicable to every airplane in every situation. Remember it and you've remembered everything that you need to deal with a dead engine on a twin. It is an all-purpose, all-around Go-away-ground procedure for immediate application whenever the earth is too near, and it works just as well if both engines are running. You can use it on takeoff, during cruise, on a go-around, or even if you've thoughtlessly descended through your clearance altitude or slipped too far below the glideslope during an approach.

It can be practiced constantly. You have heard about the practice of making believe that every twin-engine takeoff is a takeoff during which you're going to lose an engine in order to prime yourself for the possibility of trouble ahead. Great idea − as long as you have in mind a plan of action to be taken if that single-engine emergency does arise. If you don't, putting yourself on the *qui vive* can be worse than having the engine die unexpectedly.

Actually performing every takeoff as a rehearsal of full emergency procedures, however, is far better than playing mental games. Impossible without shutting one down, you say? Not really. The first step is to divide your takeoff into three segments: before Vmc, after liftoff but before running out of runway, and committed to fly.

The first is easy: If something goes wrong, pull the power off and taxi home. The second is no harder, for all you have to do if you run into trouble 50 feet in the air with a mile of runway ahead is again to pull off the power, land, and get on the brakes. If you are mentally ready to identify both of these stages of the takeoff, they present no problem − if you don't have your brain in neutral while you're rolling toward Vmc and if you don't already have the gear up and departure control on the horn while you're struggling out with plenty of runway left to come home to again.

In the third segment you literally perform the full emergency procedure: when you reach that point over the runway at which it is no longer possible to land straight ahead, go all the way through your engine-failure litany: *power up* (the mixtures, props, and throttles are already fully forward); *flaps to takeoff* (reach for the handle and confirm it, though that's doubtless where they are, unless you're using some special short-field procedure); *climb attitude* (look at the gauge — does it show climb? Do you have that best-single-engine speed nailed?); *gear up* (now do it to it); *identify* (relax, they're both running).

This same litany works for normal go-arounds, of course — another excellent time to run through the procedure.

You'll notice that the first two segments of your takeoff procedure require that the gear be down and still locked. For some reason rapidity of retraction is often equated with airmanship. The fact is, however, that leaving the gear down until it's no longer possible to land on the runway ahead is an outstanding idea, and anyone who considers moving an electric switch or a hydraulic lever from position A to position B a sign of skill and daring has a rather low skill-and-daring standard.

Most modern twins tuck their wheels away in no time but groan and crank for a somewhat longer time to get the same wheels back down again. There is a school of thought that holds that altitude is more valuable than wheels in place and that altitude is gained more rapidly with the wheels retracted, but this thought probably had its genesis back in the days when airplanes just wouldn't climb with the wheels down. Think about it: if you lose an engine at 200 feet, would you rather be able to pull power off and land straight ahead, or would you prefer to keep it flying, reenter the pattern, and make a whole new approach?

The wheels are also the prime consideration in those situations in which, due to loading and/or density altitude, you get to the climb-attitude point in your procedure — and you find that the VSI is strongly indicating "down." Unless the sink rate is very, very slight or the downward trend shows definite signs of reversal as you rotate the airplane, don't be fooled into thinking that you'll be able to turn it into climb just because you retract the gear: extended wheels actually make surprisingly little difference in rate of climb, and on some airplanes the activation of gear doors can spoil a lot of lift. If the VSI says that you're eventually going to meet the ground, the meeting should be made with the wheels down: they'll absorb shock, dispel energy, and perhaps even turn the event into a normal off-airport landing. In any case, it'll be a more pleasant event than the possible consequences of trying to climb, turn, and stagger back to the airport with what little climb you've been able to add to the airplane by folding the wheels away.

There is a fourth, less obvious, advantage of this six-step emergency procedure — of any simple, well-practiced routine: if trouble suddenly strikes, it is more important to do *something* than it is to do exactly the right thing. Activity cancels fear, and that activity can also constitute a kind of handle to grab and hang onto just at those times when the airplane threatens to run away from you.

When you're three hours into a routine four-hour flight, your chin is sinking slowly into your chest, and suddenly the bird goes BANG and tries to turn around and bite you, don't sit there and try to analyze things: say to yourself: "I have absolutely no idea what happened and even less idea what to do about it, so I'll bring the power up, flaps are clean, climb attitude, gear up, identify, and feather." You may occasionally find yourself shutting down a perfectly good engine just because a passenger nudged a fuel shutoff, but at least you'll be spring-loaded to react to the worst, and you'll never find yourself with one turning and one burning while sitting on your hands and wondering what to do.

35.

SPINS

Before the flight the pilot had said that single-engine procedures would be practiced. The instructor was an ATR pilot with over 13,000 hours, including 1,500 hours in type. The student was also an instructor − total time 1,000 hours − but with very little experience in this particular type of aircraft. Witnesses reported seeing the aircraft in level flight at about 3,500 feet and hearing abrupt power changes. One thought that the right propeller had stopped. The aircraft then nosed over into a steep right turn and spun to the right. After three or four rotations, it transitioned sharply to a flat spin and struck the ground with the wings in an approximately level attitude.

According to tower controllers on another day and in another place a single-engine aircraft appeared to be low on the base leg. As it turned final, it suddenly rolled to the left and crashed in a wing-low attitude one mile short of the runway. The pilot said later that he had overshot the turn onto final and added left rudder, up elevator, and power. The airplane rolled, and there wasn't enough altitude to recover before striking the ground.

Does all this sound familiar? It should, because, despite the efforts of manufacturers to make their airplanes controllable at stall and practically spinproof − plus pilot training in stall recognition and recovery − the stall/spin accident is still very much with us. The two previous examples could have been lifted almost verbatim from CAB accident files. Perhaps a little insight into the nature of the spin itself, as viewed by the aircraft designer, would be of considerable help as background for a better understanding of the problem.

A stalled condition is essential to a spin. Furthermore, an aircraft can be stalled in any attitude and at any airspeed within its structural limitations. This is just another way of saying that a wing stalls at a fixed angle of attack for that particular wing design, regardless of any other influences such as aircraft gross weight, G-loading, flight path, and so forth. It is not so well known however, that in the spin one wing is producing slightly more lift than the other because it is slightly less stalled. Several things at the stall can produce the roll-off on one wing that will result in a spin. Perhaps the most common is to be in a sideslip at the stall, which is indicated if the ball in the turn-and-slip indicator is not centered. Because positive dihedral effect is designed into the wings of almost all general aviation aircraft, it produces a roll in a direction opposite to the slip (that is, left ball gives right roll). Consequently, any stalled condition in which there is a yaw rate, which in turn invites some sideslip, sets up the conditions necessary for a spin to develop. In fact, the yawing motion itself could produce a spin, because the advancing wing will have a slightly smaller angle of attack and will be less completely stalled than the retreating wing.

Another way to produce a roll at the stall is to use ailerons at the break to correct a wing-low condition. This tends to introduce some sideslip because of the adverse yaw effect of the ailerons. Furthermore, deflecting an aileron downward produces an increase in the effective angle of attack of the wing, which could precipitate a stall on a wing that is already operating near its maximum angle of attack. Consequently, aileron deflection by the pilot in a direction intended to raise a low wing may actually induce roll in the opposite direction, adding to the control difficulties. Modern aileron design has largely overcome this problem. On most current aircraft ailerons are effective throughout the stall, in marked contrast to many early designs, which were quick to spin off an aileron input near the stall. This is why flight instructors say, "Keep the nose straight ahead with rudder alone near the stall!" In effect, you are attempting to keep a zero angle of sideslip − with the control best suited for the purpose.

Once the spin has begun, the aircraft is in a stable condition as long as it remains stalled. It is locked in a balance of aerodynamic and inertial forces. The aerodynamic forces induce a nose-down recovery tendency, because the airflow produces a positive angle of attack, hence a positive lift, at the elevators. But the inertial effects produce a nose-up force, much as a spinning top tends to right itself if tipped. At some nose-down attitude these aerodynamic and inertial forces balance out, and the aircraft remains in a stable spin. This description of the spin emphasizes that rotation of the aircraft about its vertical axis is necessary for the balance of inertial and aerodynamic forces that make the spin a stable condition. That is, without the rotation there are no inertia forces to cancel the aircraft's own natural recovery tendency. The spin is therefore the only condition in which an aircraft is truly stable and will not change its direction of motion if disturbed by outside forces − a fact that was occasionally used by early aviators who lacked the necessary instruments to let down through an overcast.

In theory, it is possible to predict the spin characteristics as well as the spin-recovery characteristics of an aircraft by calculations involving its mass distribution and the aerodynamic influences of its control surfaces. Because of the complexity of the aerodynamics involved in spins, however, the theory is not always completely accurate. More than one factory test pilot has ended his day's work by parting company with an uncontrollably spinning airplane. The theory does indicate which factors will most influence the spin characteristics of a particular airplane, even if these variables cannot always be assigned numerics values accurately enough to permit complete reliance on them without flight tests.

The first and most important factor is the distribution of mass in the airplane: that is, does it tend to be about the same density throughout, or is a lot of weight concentrated along one particular axis? For example, a light twin with tip tanks and wing-mounted engines would be more massive around the roll axis than around the pitch axis. Hence it would take relatively less force to change its pitch attitude than to make it roll. A reverse case would be an executive jet with tail-mounted engines. In this case most of the weight tends to be concentrated along the fuselage rather than out on the wings, resulting in a larger pitch inertia than roll inertia. The typical single-engine design has a relatively even mass distribution and, therefore no favored axis.

The importance of mass distribution arises because forces generated by different control surfaces must be able to move this mass around if the aircraft is to have proper handling qualities. For example, if you want to add tip tanks as a long-range modification, you must increase aileron power to maintain the same roll capability as before. For this reason the type of mass distribution of an aircraft will determine the control that is most effective in its spin recovery. Fortunately, in any type of aircraft, rudder is always used against the direction of the spin, regardless of the use of other controls. The vertical-stabilizer and rudder designs thus become particularly important in predicting spin-recovery characteristics. In general, the vertical area at the tail required to assure a proper spin recovery is a function of the aircraft's mass distribution and its density relative to the surrounding atmosphere. It can be expressed as:

$$\frac{(\text{mass of the aircraft/wing area} \times \text{wingspan})}{\text{air density}}$$

Relative densities vary from a high of about 35 for most executive jets to a low near 6 for a typical light single-engine trainer. In theory the rudder and fin area for adequate spin recovery is directly related to the aircraft's relative density: therefore, as the relative density increases, so must the tail size. In a sense, the tail area must be able to compensate for the lack of air resistance to rotation by small wings on a heavy aircraft, which is the case if the relative density of an airframe is high. In contrast, a glider has a low relative density and large wings that have greater resistance to rotation about the vertical axis. Gliders thus require a smaller vertical area to stop the rotation. This relative-density expression appears in many places in aircraft design, particularly in relation to aircraft stability and control calculations.

Since gross-weight increases are the natural evolution of most designs, an aircraft's relative density usually increases as newer models come out. Furthermore, optional additions such as tip tanks can also reduce the margin of acceptability in spin-recovery characteristics by adversely affecting mass distribution. This can have a great deal of significance for the original tail design, particularly for the single-engine type of aircraft, because its even mass distribution makes it highly dependent on the rudder as the primary spin-recovery control. In some instances, attempts to increase the gross weight of a design have degraded the aircraft's spin characteristics to an unacceptable degree.

For pilots the most important result of these theoretical and experimental investigations into spins is the insight that they give into the best recovery techniques for different classes and densities of aircraft. As mentioned before, the rudder is always used against the spin; the use of the other controls is not always so obvious. In the case of the typical single-engine light trainer, in which pitch and roll inertias are about the same, the standard recovery technique taught to everyone is rudder against the spin, followed by down elevator.

Since the theory works well in the area with which most people are familiar, one can move on to the more unusual cases with some degree of confidence. If the roll inertia is greater than the pitch inertia, as in the typical light twin, the spin-recovery procedure would be down elevator, followed by rudder against the spin. In the last possibility, if pitch inertia is significantly greater than roll inertia, as in the executive jets with their aft-mounted engines, the most efficient recovery is to use aileron with the spin plus rudder against the spin. In each case the control that has the least aircraft inertia to fight is the primary recovery control, which emphasizes the importance of the aircraft's mass distribution in determining spin-recovery techniques. For aircraft certificated under FAR Part 23 for non-transport-category aircraft, recovery from a one-turn spin in one additional turn with flaps up or down must be demonstrated by the manufacturer for normal-category aircraft. Or, by complying with certain center-of-gravity and elevator-travel restrictions, the aircraft must be shown to be incapable of spinning. As long as the aircraft is flown within its operating limitations, you are protected from accidental spins that leave no chance for recovery. Transport-type aircraft certificated under FAR Parts 25 and 121 have no spin requirements and are considered safe because stall characteristics are well defined and because of increased crew competence.

A few additional points should be considered, however, before you leap into your Learjet to practice some spins. In those transportation-type aircraft capable of spinning but not certificated for intentional spins, the recovery technique is likely to be quite different from that of the trainer in which you first practiced spins. If a Cessna 310 stalls, for example, it takes quite a bit more power, altitude, time, and nose-down elevator to recover than does a Cessna 150. If a spin were to occur, the nose-down elevator for recovery might very well be full forward on the wheel in spite of the already near-vertical attitude. One thing that occurs with unfortunate regularity in light twins is that a spin follows an engine shutdown. In most airplanes power has a destabilizing as well as a nose-up pitching effect. This means that any power on in a spin is going to accent the nose-up pitching moment and make it just that much harder to generate enough aerodynamic nose-down force to recover. In fact, if the power is high enough, the spin may become so flat that the elevators reach too high an angle of attack and stall. If this occurs, recovery is impossible, because there is no longer any aerodynamic nose-down force available. If a twin with only one engine develops power in a spin, there is insufficient rudder available to stop the rotation. In addition to Vmc and aft CG and Vse you would do well to know something about the mass distribution of the aircraft you fly and whether it has greater pitch or roll inertia. It could make a tremendous difference in how your flying day ends.

36.

DECLARING

AN EMERGENCY

In 1970 flight service stations alone provided 1,900 pilots with "assists." According to FAA statistics these 1,900 Pilots were in about 2,400 "situations" − one pilot may have been in two or more at once − and of these nearly 1,400 involved being lost. Over 300 involved being on top or in weather and not appropriately rated or qualified to deal with the situation, and 170 involved running low in fuel. Fewer than 350 of the assists involved mechanical or communication failures. The statistics reveal what one already knew: that most inflight trouble begins and ends with the pilot. They also reveal that the FAA is able to help pilots who get themselves into trouble; in addition to almost 1,900 FSS assists, there were nearly 1,200 tower assists and 260 air-route traffic-control center assists during 1970. (Of the total of 3,245 rescues 305 were of military aircraft.)

The FAA admits, however, that it has a problem with the pilot who gets himself into trouble and won't confess it, either because he is afraid of looking foolish to his passengers or even to the disembodied voice on the radio or else because he thinks that his difficulty may implicate him in a violation of one of the catch-all FARs about proper flight planning or familiarity with the equipment. In order to dispel at least the latter anxiety, the FAA has reduced the paperwork that follows the declaration of an emergency or the rendering of an assist to an optional report, submitted by the controller or flight-service man to the GADO, which is much more likely to lead to a helpful conversation with one of the FAA's accident-prevention-program counselors than to punitive action. Except in cases involving violations so flagrant that a hearing is unavoidable the FAA takes an indulgent position, granting virtual immunity to the pilot who admits his mistakes and is willing to discuss them and learn from them.

The mistakes are classically familiar: inadequate preflight planning and check of the airplane, lax navigation, and a kind of headstrong indifference to hints from the weather and the gas gauges. The end result is classic as well: lost on top and low on fuel. This is not the only kind of emergency, but it is one of the most common. It is a situation so familiar to many pilots, in fact, that few would even dignify it with the name of "emergency." Precisely where a tight fix leaves off and an emergency begins is impossible to say. The FAA defines an emergency as a situation in which loss of the aircraft and/or its occupants is possible or in which a forced landing may be necessary or in which ground rescue or search may be called for. The definition is to say the least imprecise: the moment an airplane leaves the ground, it can be said to be in some danger of not returning

safely. For practical purposes, however and the FAA would support this notion
— the best definition of an emergency is the broadest one. Many pilots, feeling
that declaring an emergency involves whole radio frequencies being blocked off
and 747s holding for hours on end and must therefore entail some subsequent
accounting for by the declarer, will never declare an emergency until it is too
late. All too often even a little late is too late. The first rule of emergency is to
look ahead: if it seems as though luck will be necessary to extricate you from the
present situation, it may already be taking on the dignity of an emergency. This is
the time to ask for help — not later when your fate is sealed.

The *Airman's Information Manual* contains in the air-traffic-control section
of Part I an account of the emergency procedures suggested by the FAA. The
AIM considers that "uncertainty, alert, or distress" qualify as "emergency
phases." A standard radio procedure is outlined. Except in the case of a dire
emergency, contact should initially be made on standard com frequencies. If you
were lost over the Arkansas swamplands, for example, but were within
communications range of an FSS (your omni was inoperative, say), you would
call first on 122.1 or some other FSS frequency rather than on 121.5, the
emergency frequency. There is no need to "declare an emergency": if matters
become pressing, a ground station may do that anyway. Until they get pressing,
there is no point in making dramatic announcements that have no particular
purpose.

For distress the radio call begins with *mayday* and should perhaps be made
on 121.5; for uncertainty or alert it begins with the word *pan* and should be made
on a nonemergency frequency if possible. All ground stations monitor 121.5, and
there are few places in the United States (allowing for line-of-sight limitations)
where a call on 121.5 does not bring one or two immediate replies. Few aircraft,
however, monitor 121.5: it is therefore not of much use if you are already on the
ground. It is worthwhile to occasionally test the frequency by making a call on it
while en route, including in the transmission some statement to the effect that it is
"not, repeat, not an emergency."

The radioed report, *AIM* goes on, should include the identification and type
of the aircraft; its position (or estimated position), heading, airspeed, and
altitude; the amount of fuel remaining (in hours and minutes); the nature of the
problem; the pilot's intentions; and the assistance desired. Finally, the mike
should be keyed twice for 10 seconds to permit a DF fix to be taken.

DF is, with radar, the FAA's prime method of locating lost aircraft. It is a
network of ground-based ADF stations that operate in the VHF band, permitting
a ground station to obtain a bearing from the station to any aircraft from which it
can pick up a carrier transmission. (Carrier is the basic energy transmitted by the
radio when the mike is keyed; when you talk into the mike, you add modulation
to the carrier.) A lost aircraft may be located — assuming that it is able to
communicate with ground stations at all — either by using several stations of the
"DF net" to triangulate from a single signal or by using a single station and
establishing the position of the aircraft by trigonometry. In the latter case, the DF
station asks you for a 10-second carrier signal. He obtains a bearing from this and
then instructs you to turn to a heading at a right angle to the bearing that he has
obtained. He then takes two bearings one minute apart. By dividing the angular
difference between the two bearings into 60 he obtains the time in minutes from
you to the station at the speed that you are flying. (Wind is not taken into

account, but being more or less found is much better than being completely lost.)

If loss of navigational radio prevents an aircraft in IFR flight from performing a published approach at a facility not equipped with radar, DF may also be used for an emergency IFR approach. In this case, the DF operator takes fixes with increasing frequency as the aircraft approaches the landing field. The aircraft is led over the DF site and, by a triangular or teardrop pattern, is brought into a final approach position. Because DF approaches are used only in emergencies, minimums are adjusted to suit the situation. A skilled operator can provide very accurate DF-approach guidance; in order to be skilled, however, he must practice, and most ground stations are delighted to have an opportunity to perform practice DF approaches if time and traffic permit. Since it is also useful for pilots to have practiced under VFR conditions what they may someday be called upon to perform under IFR conditions, an occasional practice DF approach is desirable. Just call an FSS and ask; they will usually be happy to oblige.

Radio failure during IFR is a nightmare that fortunately rarely gets dreamed. The FAA publishes certain standard procedures in FAR 91.127 and in part 1 of the AIM; they are logical enough but sufficiently complicated to be worth photostating and keeping in the airplane. Furthermore, two procedures that many pilots were taught in past years have been scrapped — and good riddance. It used to be that if, for some reason — such as simultaneous loss of com and nav due to a complete electrical failure — it was impossible to comply with the rules outlined in 91.127, the pilot was told to fall back on two devices that dated from a remote time when controllers kept the radar gain turned up because there were so few targets on the scope. One was dropping "boxes of chaff" in a specified pattern. Most of us have never even *seen* a box of chaff, let alone been accompanied by one aboard an airplane. The other, which on the face of it seemed more plausible but turned out to be equally useless, was flying triangular patterns. Left turns meant one thing; right turns another. No one noticed. In fact, aviation writer Don Downie, suspecting that a nontransponding target flying triangular patterns might not be recognized by a controller at all, tried getting help — or at least some attention — by flying what came to seem like eternal triangles. His suspicions were verified. Fortunately, it was only an experiment he was flying, not an emergency. On a brighter and more realistic side, there is a transponder squawk — 7700 — for emergencies and another — 7600 — for the specific emergency of "lost communications capability." Radar-scopes are set to give off an alarm signal whenever they pick up a 7700 or 7600 squawk; incidentally, you have to be careful to avoid switching *through* one of these two codes while on the way between two others.

Besides the published procedures there are several practices that are valuable in certain emergencies. One is knowing the best-range and best-endurance speeds of your aircraft and knowing when to use which. Best-endurance speed is the speed at which the airplane can remain airborne for the longest time on a given amount of fuel and is about halfway between best-angle-of-climb and best-rate-of-climb speeds. It would be used if you were lost and attempting to make radio contact with the ground, holding because of dubious weather, flying triangles, or at any time that the chief aim is simply to *delay* the ending of the flight. If, on the other hand, you want to get somewhere, use the best-range speed, which is about 40% higher than the best-endurance speed. Best-range speed will take the airplane the longest distance on its remaining fuel but will not keep it aloft so long as the best-endurance speed; it is therefore useful only if you know where you are going.

The FAA recommends climbing if you are lost, for better radio communication and easier DF fixes and also because it may bring into view landmarks that were invisible at a lower altitude. If you are running low on fuel, run the tanks entirely dry — except on a fuel-injected engine or in any other situation in which engine restart is uncertain — but keep one tank a quarter full to use last. If a forced landing appears even remotely likely, begin to check the ground for indications of the wind direction and strength. If a forced landing is inevitable, make it a precautionary landing by finding the most suitable place and landing before fuel starvation occurs. If you are forced down by weather, you would do better to land on a road or in a field or meadow and wait it out with some fuel aboard than to press on beyond your capabilities. Off-airport landings are a bizarre experience for most pilots, of course, but there are a few situations in which they are the only wise solution.

The most important thing to remember about any emergency is that, except for those involving sudden mechanical failure, they develop more or less gradually. You have fair warning that things are not going well, but it takes a grown man to ask for help, and the most dangerous of pilots is the one who won't admit a mistake. Unfortunately airplanes — like cars — bring out the infant in their pilots: many will cross their fingers and take completely unnecessary risks to avoid calling somebody to say that they've gotten themselves into trouble. There is the humiliation and the presumption that admitting to difficulties puts one in line for all sorts of written reports and possible punitive action from the FAA. Humiliation is salutary, however, and the paperwork is simply imaginary. What paperwork may be done is done by the FAA, and a cooperative pilot runs no risk of being punished for asking for help. Remember that human performance deteriorates as tension rises, and panic can reduce an otherwise competent pilot to the point at which he has to have step-by-step instructions to operate his omni receiver. Fear can turn any complication into a emergency: don't wait for it to develop. If you are uncertain, call a ground station and *say* that you're uncertain: you won't be declaring an emergency — you'll be avoiding one.

37.

FORCED LANDINGS

The inflight emergency is introduced to the student pilot in a widely accepted, time-worn manner, establishing a routine that remains constant throughout his training. The instructor abruptly closes the throttle and declares a power failure. The solution to this problem requires that the student maintain aircraft control, establish the proper glide speed, select a suitable landing site, and plan an approach. The student is also instructed to troubleshoot the cause of power loss if time allows by use of carburetor heat, mixture, fuel selector, and anything else that is conveniently available. The student knows that these attempts will not correct the power loss and is also aware that his real task is to find the landing site that his instructor has already selected. He must then execute a reasonably safe approach to this field.

In reality this procedure is nothing more than an exercise in judging glide distance and sink rate and in maneuvering at a constant glide speed. Unfortunately, it stresses two assumptions that are rarely compatible during an actual forced-landing encounter. The first myth is that a suitable landing area is always available; the second, that power failure is the primary reason for attempting an emergency landing.

During training the pilot is taught to utilize only those emergency fields that can be considered "suitable landing sites." This factor is stressed to such a degree that most pilots will not even consider a precautionary landing unless they are certain that no damage will result to the aircraft. Because of this attitude, many fatal crashes attributed to "continued VFR flight into IFR conditions," or "spatial disorientation," are actually a direct result of the pilot's desperate attempt to continue flight because the terrain below did not meet his ingrained standards as an emergency field.

The modern aircraft is so reliable that forced landings due to mechanical failure are extremely rare. Most forced landings are the result of pilot errors such as poor flight planning, fuel mismanagement, or proceeding into marginal weather. A recent NTSB accident summary contains information on 898 general-aviation accidents processed during the two months preceding the report. "Continued VFR flight into adverse weather conditions" caused a total of 27 accidents, of which 17 were fatal. There were 19 crashes caused by "spatial disorientation," and 16 of these were fatal. It would seem that the chances of surviving such accidents are remote. "Selected unsuitable terrain" was the cause of 60 more accidents — but of this total only one crash was fatal. It is obvious which is the lesser, more survivable evil. Before most fatal crashes occur, there is usually a moment at which the pilot can choose the type of accident in which he will be involved. Before a pilot continues into deteriorating weather and becomes trapped under or in the clouds, he has the opportunity to make a precautionary landing. All too often this alternative is rejected, and he accepts a deadly option: flying into the ground or losing control due to spatial disorientation. To increase his chances of survival, the pilot should realize that *almost any terrain can be considered suitable for a survival crash landing.*

Psychological hazards often influence a pilot's ability to act promptly and decisively in an impending emergency. His mind becomes paralyzed at the thought of his self-inflicted predicament, and he submits to an unconscious desire to delay any irrevocable action as long as possible. His training has conditioned him to expect to find a suitable landing area, so he presses on, hopeful that one will appear. The alternative to this situation can be a crash landing on poor terrain. This choice is usually not considered, yet there are proven techniques that involve use of the aircraft structure to protect the occupants during a touchdown on poor terrain or in trees. First, however, a pilot must overcome his intense desire to save the aircraft at all costs. A pilot will often ignore every basic rule of airmanship to avoid landing in an area where aircraft damage is unavoidable. He'll make a steep 180 at low altitude back to a runway rather than crash-landing straight ahead, or he'll stretch a glide to reach a more attractive landing site. A pilot must also overcome his certain belief that a damaged aircraft equals bodily harm. The success of a crash landing depends as much on mental attitude as on flying skill.

Once the decision has been made to attempt a landing on terrain on which substantial damage to the aircraft will result, the most important considerations are the aircraft's speed at touchdown and its rate of descent.

The severity of deceleration will be directly related to the groundspeed at touchdown, as will the stopping distance. Double the speed and you quadruple the total destructive energy. An impact at 85 knots is twice as violent as one at 60 knots; touchdown at just over 100 knots is three times as dangerous as at 60 knots. Touchdown must obviously be made at the lowest possible *controllable* airspeed, using flaps or other aerodynamic devices.

While most pilots will choose the largest available landing site, very little stopping distance will be required if the speed is uniformly dissipated over the available distance. The average general aviation aircraft is supposed to protect the occupants in crash landings that expose them to nine times the acceleration of gravity in a forward direction, and at 45 knots a uniform nine-G deceleration requires only *9.4 feet* to stop an aircraft. At 87 knots the distance required is four times as long, or 37.6 feet. The key to these short stopping distances is uniform deceleration. The pilot can take advantage of dispensable structure on his aircraft in landing on poor terrain and accomplish this deceleration over a short distance, reducing peak deceleration of the cabin/cockpit area.

Since the body's tolerance to vertical Gs is extremely limited, a high sink rate and hard touchdown should be avoided even in landing on ideal terrain. If the touchdown is made on rough or soft ground, a hard landing may cause the aircraft to dig in, resulting in severe forward deceleration or flipping over. An excessive nose-low attitude should be avoided for the same reasons. Once he is committed to touchdown on rugged terrain or in heavy brush and trees, the pilot must keep in mind that avoiding injury depends on his ability to keep the vital structure of the cabin/cockpit area intact. This can be accomplished by sacrificing the wings, landing gear, fuselage bottom, and other dispensable structures. Another energy-absorbing medium may be available at the landing site. Trees, vegetation, dense crops, or even man-made structures such as fences can be effective arresting devices to bring an aircraft to a stop with repairable damage.

It cannot be stressed strongly enough, however, that all of the energy-absorbing and decelerative material in the world will be of little use if the cabin stops and its occupants keep going. Seat belts and effective shoulder harnesses are critical in preventing injurious contact with the interior structure, and the efficiency of harnesses has been proven so often that it's a wonder that there are still pilots around who refuse to use them.

Since the aircraft's speed at touchdown is of prime importance, the use of full flaps is recommended, but caution should be exercised in extending them to avoid premature dissipation of altitude and speed. Although many steps may be taken to configure the aircraft in order to reduce other hazards associated with a crash landing, positive aircraft control must take priority over all other considerations. There is no definite rule on the position of retractable landing gear. On rough ground or in trees extended gear would offer protection, but this advantage must be weighed against the possibility of rupturing fuel tanks. On soft terrain or plowed fields a gear-up landing may result in less damage than one made with the wheels down.

Use available power to assure a good approach, but, after the field is made, switching off the mags and fuel can appreciably reduce the chances of post impact fire. A cool engine also lessens the hazard of fire.

Just as in the standard forced-landing technique, there are three important factors to consider while maneuvering for the approach: wind direction and velocity, obstacles in the approach path, and landing-area size and slope. Since these factors will rarely be as desired, you should select a compromise that best allows for errors in judgment. Trying to stretch the glide over trees or powerlines may prove less advisable than choosing a clear approach in a crosswind or even a downwind. Collision with an object at the end of the ground roll (or slide) is far less hazardous than hitting one while you are still airborne.

If you are landing in an extremely confined area, plan your touchdown *prior* to impact with trees. An aircraft decelerates much faster on the ground than in the air. If necessary, force it on rather than waiting for it to settle. Although you will find textbooks that advocate flying between two trees as a means of deceleration, never contact an object while you are airborne unless it is absolutely necessary. The touchdown should be made in low, heavy brush or small trees. In this instance, make contact when the aircraft is slightly nose-high while still at controllable airspeed. The foliage will provide a cushioning effect for the settling aircraft while helping to decelerate it at a uniform rate. Being forced down in heavy forest may prove terrifying, but such an event is survivable if a few general rules are followed. Select the lowest possible stand of heavy-crowned trees of equal height. Landing in tall trees is undesirable and may result in a destructive free fall to the ground after the aircraft has stopped. Use the recommended landing configuration for your aircraft, gear and flaps down into the wind. At impact the aircraft should make contact at minimum flying speed in a nose-high attitude. Don't stall but "hang" the aircraft in the tree crowns, involving both wings and the fuselage bottom simultaneously. This will provide the cushioning and even deceleration desired while preventing violation of the windshield.

Landing on open mountain slopes may look more inviting but often requires a higher degree of skill. Slope landings should always be made upslope if possible, and if you are landing on a pronounced upslope, maintain sufficient airspeed to allow a considerable change in pitch attitude just prior to impact. Note carefully the difference between your glide angle and the upgrade of the slope. Remember that you will be changing abruptly from a glide to a climb just at touchdown. A normal descent of 500 feet per minute at 45 knots equals a glide path of 6 degrees. If the slope that you are approaching is 24 degrees, the resulting change in your pitch angle will be at least 30 degrees at the time of flare.

Since each emergency-landing situation involves specific problems with many variables, no precise rule can be stated regarding the pattern to be flown. Sometimes there is insufficient time even to consider one. The most important consideration is to reach your intended landing spot by normal flying techniques − slipping, S-turns, key position and speed control.

A pilot's choice of landing sites is often governed by actions taken in preflight planning. The route selected and his height above the terrain when the emergency occurs can broaden or severely limit his options. The only time that a pilot has no choice − or at best a very limited one − is during the low/slow period of takeoff. Even in this event, however, use of the techniques described may help him to realize that by changing the impact heading only a few degrees he can greatly increase the chances of a survivable crash.

Should the emergency begin at a considerable altitude, it is wise to first select only the general area rather than a specific field. From altitude ground contours can be misleading, and you may lose the greater portion of your altitude placing your aircraft in the wrong approach position. Wait until you can be fairly certain of the terrain before committing yourself to a field, and even then don't hesitate to change your mind in favor of another that is obviously more acceptable − but don't change your mind too often or at a low altitude. A properly executed landing on poor ground is more desirable than an uncontrolled crash on a paved airport. A precautionary landing is always less hazardous than a forced landing, so don't hesitate to bend some aluminum if the alternative is breaking your neck.

38.

GLIDE, BABY, GLIDE

Losing your engine at altitude need not be a nightmare. Emergency landings can be safe and easy if you know your plane's best power-off gliding speed and how far you can glide at that speed, for then you can determine your radius of action power-off or your minimum cruising altitude for various situations. Although this information is vital for all pilots — especially single-engine pilots who fly over metropolitan areas or rough terrain — few owner's manuals provide gliding information. Through a few simple measurements and calculations, however, this information can be found, and you can spend an interesting afternoon doing the job in flight.

The gliding performance of a falling object is called the glide ratio. The glide ratio always equals the lift of a body divided by the drag of the body, which is also known as the lift-to-drag ratio (L/D). The L/D or glide ratio expresses the distance a body can travel horizontally for every unit of vertical height lost. For example, an "aerodynamically dirty" airplane may have an L/D ratio as low as 6 to 1. This means that the plane will glide 6,000 feet forward for every 1,000 feet of altitude lost. Modern jet transports have glide ratios of about 14 to 1, while high-performance sailplanes claim to have L/D ratios of up to 50 to 1. Every airplane, no matter how high its wing loading, can be operated as a glider. NASA research pilots such as Neil Armstrong made power-off approaches in the North American X-15, even though its vertical rate of descent exceeded the horizontal-landing-approach speeds of most lightplanes.

What about the airplane that you fly? Does the owner's manual provide glide information? If it does, all you have to do is memorize the best glide speed and the glide ratio, but, if the information is unavailable for the airplane you fly, there is a simple flight-test procedure that will provide it. Since glide ratio or L/D equals horizontal speed divided by vertical speed, the problem is one of finding the forward speed that gives the greatest L/D ratio, or the farthest glide. For flight testing choose a reasonable speed range around which you think your plane will glide the farthest — between 60 and 85 mph for a typical fixed-gear single, say. Prepare a separate 3" x 5" card for each 5-mph increment within the chosen span so that your copilot can record data while you fly.

Pick a calm day for the test so that gusts and turbulence will not affect the results. Choose a safe altitude and area and begin a glide by idling the engine(s) with the airplane in its cleanest configuration. Maintain wings level and obtain the chosen speed with the elevators. Once you establish a steady glide — and only then — have your copilot start the stopwatch or note the time and record the altitude. At the end of three minutes, have him note the altitude again. Take the outside air temperature before and after one of the glides so that you will have a value to use in computing true airspeed later on. After each run it is a good idea to fly at moderate power for a few minutes to warm the engine(s) before adding power to climb back up for the next glide: you will find that the cylinder-head temperatures drop considerably during the glides.

The engine(s) should ideally be set up to give zero thrust. Owner's manuals for some aircraft equipped with constant-speed props have this information, but if zero-thrust settings are not known, use a constant low idling speed throughout the tests. In the case of fixed-pitch props an extra-slow idle created by the addition of full carburetor heat will simulate the drag caused by this prop in a power-off situation.

Use your data to make a "glide polar curve" chart that will show your plane's rate of descent at all speeds. For each speed tested find the altitude lost between the start and finish of the run and divide this by the number of minutes spent in the glide. This will give your rate of descent in feet per minute. Plot the rate of descent versus the gliding test speed (corrected to TAS using the average OAT and altitude recorded) on a graph and connect these points with a smooth curve. This is the glide polar curve. To find the speed that gives the maximum glide distance, simply draw a line from the zero velocity and zero rate-of-descent point at the origin of the axes of the graph to the point of tangency on the curve. Draw another line straight up from this point to obtain the magic speed. To calculate the best glide ratio, convert the mph to fpm by multiplying by 88, since 1 mph equals 88 fpm. (If gliding speed is measured in knots, multiply by 101.3.) Then, use the numbers that you've obtained and the formula "glide ratio equals horizontal speed divided by vertical speed." For the sample airplane the formula reads:

$$\text{maximum glide ratio} = \frac{72.5 \text{ mph x } 88}{785 \text{ fpm}} = 8.1$$

The airplane will therefore travel 8.1 miles forward in still air for every mile that it sinks.

No other speed-and-descent combination will give an engine-out glide as far as the one found by this method. Even though the nose may appear low at the best glide speed, pulling it up to a more comfortable-looking angle will actually shorten the distance traveled and lessen the chances of making the runway or the emergency-landing site.

Surprisingly enough, the speed that gives a minimum rate of descent is less than that for maximum gliding distance. The minimum sink rate for the sample airplane is 752 fpm and it occurs at 66 mph, but it gives a glide ratio of only 7.7 to 1 compared to 8.1 to 1 at the best glide speed. At 66 mph the plane is losing altitude at a slower rate than at the best glide angle. Though this delays the forced landing, the airplane will not have traveled as far over the ground by the time that inevitable landing is made.

Admittedly, using the slant-distance gliding speed taken from the airspeed indicator instead of the true horizontal speed across the ground will introduce error in the L/D calculation. For airplanes with glide ratios greater than 7 to 1, however, the glide angle will be less than 8 degrees below the horizon, making this error less than 1 percent.

To get the most out of your plane's glide ratio, be aware of whether or not changes in gross weight, altitude, and winds aloft affect best-glide speed and distance traveled. Gross weight has no effect on glide distance if adjustments are made in the gliding speed: a lead Cherokee would glide just as far as a balsa-wood Cherokee. If your plane is heavy, however, glide a little faster than the maximum L/D speed; if light, a little slower. The maximum gross weight and minimum flying weight of general aviation airplanes usually won't differ by more than 40% of the maximum, meaning that glide-speed variation between weight extremes will be less than 25%. (Sailplane pilots sometimes use this phenomenon to their advantage by adding ballast to their ships for cross-country competitions in order to boost their best glide speed and finish the race quicker. Of course, they suffer the penalty of a higher rate of descent, but this just means that they have less time to find the next thermal.)

Altitude also has no effect on glide ratio, provided that a constant indicated airspeed is held throughout the descent. Although the true airspeed will be decreasing as altitude is lost, the glide ratio will remain constant in a no-wind condition. If you find the maximum L/D of your plane to be 11 to 1 in tests conducted at 10,000 feet, your best glide ratio will still be 11 to 1 at sea level as long as the same indicated airspeed is maintained. Because the true airspeed is higher at altitude, the vertical speed will also be proportionally greater, so don't panic if you see a large rate of descent up high.

Winds obviously affect gliding distance, so fly a little faster than the maximum L/D speed if you are gliding into a known headwind – to reduce the time during which the wind can hold the airplane back. Conversely, hold a slightly slower gliding speed in a tailwind to take advantage of the wind's effect on distance traveled. If this trick sounds a bit phony to you, think of an airplane with a best-glide speed of 60 mph gliding into a 60-mph headwind. Holding best-glide speed will get him nowhere: he must glide faster.

If angle-of-attack indicators were standard equipment on light airplanes, there would be little reason to publish this chapter, since there is one fixed angle of attack at which a given airplane will glide the farthest in its clean configuration. The optimum angle of attack is independent of all atmospheric or aircraft loading changes except wind correction. Since pilots today are forced to fly by indicated airspeed alone, the problem becomes one of finding the speed that corresponds to the angle of attack for maximum L/D ratio. But wouldn't it be easier if we all had a meter with a line painted at the optimum angle of attack?

39.

THE STRUCTURE
OF AIRSPACE

Publisher's Note: The material in this chapter is not only obsolete, but paying it any heed except as a curiosity item would certainly be dangerous, thanks to changes in Federal Regulations. It is included only for the sake of completeness and a sense of responsibility to re-publish this book in its entirety.

The airspace structure is just a little more complicated than an unassembled Christmas toy. Figuring out how it's put together would be easy if the part names matched their purpose, but they all sound alike: think about any two of them and you're bound to become confused.

Control zone and airport traffic area, for example: airport-traffic areas exist in every control zone where there is a tower. Why doesn't every control zone have an airport? Why doesn't every control zone have a tower?

Continental control area and positive control area occupy much the same airspace, so you might wonder why there are two of them. You feel safe in assuming that each is a different form of control area, but unaccountably *neither* is a control area. (Is a terminal control area a control area? Alas, a TCA isn't a control area, either.)

Whatever it is, continental control area seems pretty big. Positive control area is the "airspace within which there is positive control of aircraft." Good Lord, how is traffic controlled elsewhere?

Although CCA, PCA and TCA are not control areas, they are all different forms of controlled airspace.

Control area is a nondescript term representing some but by no means all, forms of en-route controlled airspace. It used to be the generic term for whatever airspace was under ARTC jurisdiction. Control zone delineated control-tower operators' airspace. All the uncontrolled airspace was nameless except if controllers referred to it as elsewhere.

Tower-controlled airspace hasn't grown much − not at all, in fact, until TCAs were invented − but center expansion, encouraged by the ubiquitous security blanket of radar, has exceeded Caesar's grasp. Centers have delegated much of their low-level airspace to towers for approach-control service.

Some forms of airspace overlap each other, some butt up against each other, some carve chunks out of others, some are based on weather minimums and others on equipment limitations, one is a speed zone, and some have varying dimensions − expanding and contracting in reaction to their distance from dimensions navaids.

The best way to sort these out is to look at them all, the common ones and the rare ones. Perhaps especially the rare ones, because they are the ones that confuse us. Very few pilots, for example, need to know anything about control-area extensions. Anyone who first hears the term, however, might wonder what it is and why it is and to what extent it interferes with his flying.

UNCONTROLLED AIRSPACE

In the beginning the air was pristine and uncontrolled from bottom to top. Now, only a portion is uncontrolled, and it has three layers: from the surface to 1,200 feet agl; more than 1,200 feet agl but less than 10,000 msl; more than 1,200 feet agl and at and above 10,000 msl. (The three layers are formed by three different sets of VFR minimums.)

The 1,200-agl layer of controlled airspace is more than coincidental with the 1,200-agl floor of most airways. VFR pilots can scoot across the blue lines on sectional charts by remaining clear of clouds and keeping one mile of visibility if they fly less than 1,200 feet above the surface. No one recommends a three-hour cross-country in such conditions, but it's not hard to visualize the many times when these minimums in nonmountainous terrain above 10,000 feet msl came about because of the rapid closure rate of the faster aircraft flying at higher altitudes. Over the 48 states and most of Alaska uncontrolled airspace (the shaded portion of IFR en-route charts) tops out just below 14,500 feet msl, where the Continental Control Area begins. At that altitude, however, the same VFR minimums apply whether flight is in uncontrolled airspace or within the Continental Control Area.

SPECIAL-USE AIRSPACE

There are just two kinds of *special-use airspace*, prohibited areas and restricted areas. Other airspace reservations, which the public has no part in establishing, will be considered separately. *Prohibited Areas* are places to avoid. There are about half a dozen of them, all more or less related to security or national welfare. They begin at the surface but extend vertically only as high as necessary. At Mount Vernon the upper limit is only 1,500 feet msl. The highest are at Presidential locations, like Washington, D.C., where the tops are at 18,000 feet msl.

Restricted Area: airspace hazardous to fly through. There are used for special maneuvers by military aircraft, for test artillery and antiaircraft shells and the like, and for keeping nonparticipating aircraft a proper distance away from sensitive ground installations. Their size is strictly limited to the minimum needed. Many are joint-use restricted areas, which means that the public can fly through if they are not in use for the set-up purpose. The nearest ARTC center knows when they aren't in use.

CONTROLLED AIRSPACE

Controlled airspace is under ATC jurisdiction. It is established after the public has been notified of the government's intent and has had an opportunity to respond.

The *Control Zone* is the only form of controlled airspace that extends down to the surface. Control zones provide empty air during poor weather conditions at airports where a pilot on an instrument approach has enough to contend with without worrying about whether he'll encounter local traffic hidden behind low scud in the pattern after he breaks out. Control zones are also established at airports with a large volume of VFR traffic, whether or not an instrument-approach procedure serves the field. VFR traffic in control zones without a tower is uncontrolled during good weather. During weather below basic VFR conditions the zone is under the jurisdiction of the center (or approach control) even if there is a tower in operation. In other words, tower operators control traffic only during VFR weather conditions. Special VFR clearances (in which VFR aircraft are given IFR separation within the zone) come from the center or approach control. Control zones have at least a 3-mile radius around the reference point of the controlled airport. Most control zones are 5 miles in radius plus extensions, but they vary. Atlantic City's is 12 miles; at Bismarck, North Dakota it's 5 1/2 miles. Control zones extend outward along instrument-approach courses to the point at which arriving IFR aircraft descend below 1,000 feet agl.

The upper limit of control zones is the floor of the Continental Control Area. Since there is no Continental Control Area in Hawaii or the Alaska peninsula, control zones in those regions have no upper limit.

There are no radio-communications requirements for flight within a control zone, and clearance to enter is not needed if cloud distance and cockpit visibility meet basic VFR requirements. Flight below the ceiling without a clearance is not allowed, however, if the ceiling is below 1,000 feet. Even if flight visibility is good, no one may enter the traffic pattern without a clearance if the reported ground visibility is less than 3 miles. A weather observer must be on duty to take hourly surface and special weather observations.

The *Airport-Traffic* Area is a speed zone in which radio communications with towers is required. The speed requirements are now part of a larger ruling on airspeed, which obscures the purpose of the airport-traffic area. The maximum speed limit is high enough (156 knots for reciprocating engines, 200 knots for turbines) so that most lightplane pilots can comply with requirements merely by communicating with the tower prior to entering. Communication (or a written agreement not to) is required even if you are operating at a small field adjacent to the controlled airport. Airport traffic areas exist only in control zones in which there is a tower and only during the hours that the tower is in operation. With but one exception (at Anchorage, Alaska, where the airport-traffic area is large and complicated) every airport traffic area has the same dimensions: a 5-mile radius from the center of the controlled airport extending upward to but not including 3,000 feet agl. Airport traffic areas do not occupy control-zone extensions beyond the 5-mile radius. Some airport traffic areas are larger than the control zones on which they are centered: Meigs Field, Chicago has a 3-mile control zone and a tower. The same speed and communications requirements exist at Meigs as at any controlled airport with a 5-mile control zone.

Terminal Control Area: for controlling VFR traffic beyond the control-zone boundary and mixing it with IFR traffic for a smooth overall flow. TCAs thus far have been set up at 22 locations. The airspace dimensions vary with location and are so complicated that special charts are required. There are three types, or groups, with varying requirements, those for Group I being the most stringent.

Within Group I TCAs a transponder with encoding altimeter is required; student pilots are prohibited from operating at the principal airport; and *all* aircraft are separated from each other. This means that an ATC clearance is required to enter and that vectors, holding instructions, altitude assignment, sequencing, and all the other standard IFR procedures are applied to VFR traffic. Group II differs from Group I mainly in that student pilots are not excluded and an encoder is not required. Group III TCAs merely require that pilots *either* maintain two-way communication with ATC *or* use a transponder in the altitude-reporting mode. Basic VFR minimums apply within TCAS, and, since the airspace below the floor is transition area, basic VFR minimums apply there as well.

Transition Area: a gap-filler in the ATC system, inserted where IFR aircraft are not protected by other controlled airspace. For example, if an instrument-approach procedure required that an aircraft descend below the airway prior to reaching the control-zone boundary (extension or otherwise), a transition area would be established to keep the aircraft within an envelope of controlled airspace while keeping the control-zone extension as short as possible. Transition areas are similarly used for departure, holding, and en-route operations. If they are connected with instrument-approach airspace, transition areas have a floor of 700 feet agl. For en-route operations the floor is no lower than what is needed for IFR traffic (a portion of the Boston transition area has a floor at FL 200, for instance) and in no case lower than 1,200 agl. If it is underlying an airway, the floor of the transition area and of the airway are coincident. Transition areas therefore underlie only those segments of airways that have a floor *above* 1,200 feet agl. If a jet route does not directly overlie an airway, a transition area will be established between them, unless one is already established to match radar coverage. Radar scans huge cylinders of airspace, but ATC is limited to operating in controlled space. Transitions give controllers the airspace that they need to control the targets that they can see. Entire states, particularly in the East, are covered by transition areas, generally from 1,200 feet agl up.

Airway: the original en-route controlled airspace. Most airways are formed by VORS, but there are still a few low-/medium-frequency colored airways left. Amber 10 cuts across eastern Maine for about 9 miles; Blue 19 sits between Key West and Perrine, Florida; Alaska has a few colored airways.

All 50 states have VOR airways. Like U.S. highways, VOR airways have even numbers if they are oriented east-west, odd if north-south. Airways extend offshore to the domestic/oceanic control-area boundary. They do not encroach on prohibited areas, although they cut right through joint-use restricted areas. Airways generally have a floor at 1,200 feet agl, except if they are raised by a higher-level transition area or offshore, where the floor generally is 2,000 feet above the surface. Incidentally, a modicum of vertical separation exists between the lowest IFR aircraft on an airway and any VFR aircraft passing underneath and flying clear of clouds. MEAs are several hundred feet (usually 500 or more and never less than 300 feet) above the floor of the airway. The lowest MEA you will find is 1,500 feet msl. The upper limit of airways touches the floor of the jet-route structure, which begins at 18,000 feet msl, except in Hawaii, where airways have no upper limit. Airway segments offshore have no upper limit. The standard width of airways is 4 nautical miles on either side of the centerline, expanding at a 4.5-degree angle beginning 51 miles from the navaid because of decreasing accuracy of the received radio signal. There are a few exceptions, such as V2 between Grand Rapids and Lansing, Michigan, where the airway is 7 miles wide − 3 miles north and 4 miles south of centerline. Victor airways have alternate airways laterally spaced at least 15 degrees apart. The triangular chunk of airspace between the main and the alternate segment is usually established as a control area.

Area Low Route: routes established for those who disagree with the government's placement of navaids. Inertial platforms, doppler radar, and course-line computers enable pilots to get straight-line course guidance without having to overfly navaids. Inertial guidance systems and doppler radar, in fact, don't even use VOR/DME navaids. Area low routes have a few things in common with airways and differ from them in some important respects. They have the same floor and ceiling, and each has a basic width of 4 miles on either side of centerline. Transition areas affect area low routes just as they do airways. Area low routes cut through joint-use restricted areas just as airways do. Unlike airways, which may have gaps in signal coverage at MEA, all approved area low routes have uninterrupted VOR/DME signal reception at all altitudes.

DME slant range doesn't have any significant effect on the accuracy of airways but complicates RNAV routes considerably. Unless an aircraft is using RNAV equipment that automatically corrects for slant-range error (not all do), it will tend to be pulled in toward any vortac selected to form the route of flight as it reaches a point tangent (perpendicular) to the station. Along routes and at altitudes at which the error is significant route width is expanded on the side toward the vortac. The amount of expansion is equal to accumulated error, which is a combination of altitude, distance between the selected route and the vortac, and allowable errors in the ground and airborne equipment. As with airways, area low routes (and the protected airspace of area high routes as well) are expanded on both sides of centerline to compensate for degradation of radio signal. Expansion (at a 3.25-degree angle) depends not only on the distance along centerline from the navaid but also on the distance that the vortac is offset from the centerline. RNAV routes are narrowest (except where the pulling-in effect applies) where the centerline is tangent to the station. Increase the offset distance of the vortac or the distance along the centerline from the tangent point, and the route may have to be expanded as much as 16.61 miles each side of centerline. Route width is important to pilots flying IFR on routes parallel or nearly so to aircraft on other routes. RNAV equipment enables a pilot to custom-build his own routes. In doing so slant-range error, reception altitudes, and obstruction-clearance altitudes are unknown factors: therefore custom-built IFR-RNAV routes are monitored by ATC radar.

Continental Control Area: originally an area of higher VFR minimums because the high altitudes are occupied by faster aircraft. Those same minimums now apply everywhere more than 1,200 feet agl and above 10,000 feet msl.

The CCA covers the 48 contiguous states and Alaska west to 160 degrees west longitude. Hawaii has no continental control area. The floor is 14,500 feet msl exclusive of the airspace less than 1,500 feet agl. Also excluded are prohibited areas and non-joint-use restricted areas. The CCA has no upper limit: it extends as high as the highest molecule of air. Although it is controlled airspace in its own right, most of the IFR traffic within the CCA is controlled, because some other form of controlled airspace is embedded within: airway, area low route, transition area below 18,000 feet; transition area or PCA at 18,000 feet and above.

Unlike airways and area low routes in the "low-altitude" structure, jet routes and area high routes, which are no wider than the centerline of an airplane, are not really types of airspace. Each jet route and area high route, however, has protected airspace on both sides of centerline. Protected airspace within the CCA is intrinsically controlled airspace. The ceiling of jet routes and area high routes is FL 450. Above that altitude routes of flight are direct between navaids not more than 200 miles apart. Above FL 600 navaids can be any distance apart.

Positive Control Area: high-altitude controlled airspace in which IFR flight is obligatory and in which aircraft must be equipped with a transponder − the most restrictive of all forms of controlled airspace. ATC controls everything that moves. VFR flight of any sort is prohibited − even VFR conditions on top in lieu of a "hard" altitude. This is Big Brother's airspace, ultra-right-wing-conservative airspace. PCA has radar coverage throughout, is limited to the 48 contiguous states, and exists entirely within the continental-control area. The floor is now 18,000 feet msl, and the ceiling FL 600. (The new aircraft flying above 60,000 feet fly on direct routes and are separated vertically by 5,000 feet and laterally by 25 miles each side of centerline.)

There's nothing very complicated about PCA. Its existence, however, poses a threat to the non-IFR-rated pilot. Established experimentally in the Indianapolis-Chicago area in 1960, it was bounded by the area scanned by three radars and limited to the altitudes between FL 240 and FL 350. When it was implemented nationally, its floor was set at FL 240. If the VFR pilot is ever to lose his privileges, the blow will probably occur as a result of the PCA curtain as it comes down. Before that happens, ATC will have to find a way to control all the Super Cruisers and Cessna 120s winging from pea patch to alfalfa acre. That's a long way off.

Control Areas and Control-Area Extensions: like the *Compleat Training Course* ("if it wasn't covered in the lectures, it'll be covered in the test"), are kinds of airspace that don't package neatly into the other categories. These control areas are associated with jet routes outside the Continental Control Area (across the Gulf from New Orleans to the west coast of Florida, for instance). Their vertical dimensions match the route (18,000 msl to FL 450). Their lateral dimensions are the same as an airway (4 miles each side of centerline, expanding 51 miles from the navaid).

Among the *additional control areas* (that's what they are called) the most noticeable are the *offshore-control-areas*, with numbers such as Control 1141, Control 1177, and so forth. They connect the domestic navigation/ATC system with the international: domestic ATC procedures are used. The floor generally 2,000 feet above the surface; there is no upper limit and no centerline. The lateral boundaries are formed by tangent lines connecting circles at each end. The circle at the ocean end is usually larger, which gives the routes their funnel shape. They terminate at the oceanic-control boundary.

Another kind of additional control area is the direct route you sometimes encounter in the domestic low-altitude structure. The route between the Ottumwa, Iowa vortac and the Kansas City vortac is one example.

Control-area extensions complicate few lives. There are only three of them − two in Alaska and one at Eniwetok Island in the Pacific. Control-area extensions are for controlling terminal IFR traffic that can't be kept within the confines of an airway or control zone — sort of an IFR/TCA.

NONRULEMAKING AIRSPACE

In this category fits all the airspace set up without the consent of the governed, either because compliance isn't obligatory or because the airspace is outside United States jurisdiction or because circumstances require immediate airspace action.

Terminal Radar Service Area: junior sibling of the TCA except that participation in the TRSA's Stage Ill radar service is voluntary. TRSAs are not shown on most VFR or IFR charts, and, unless a pilot desires radar following, he is free to ignore them.

Alert Area: the old definition of a caution area, which alert areas have replaced: airspace in which a visible hazard to flight exists. Alert areas typically have a high volume of military pilot-training activity, such as that at the Army's primary helicopter school near Mineral Wells, Texas. Neither waivers nor exclusions are provided; no flight restrictions or communications requirements exist. Pilots of both participating and transiting aircraft are equally responsible for collision avoidance.

Intensive Student Jet Training Area: airspace in which student jet pilots accumulate experience. The floor is set at the highest possible altitude, never less than 8,000 feet msl or 4,000 agl, whichever is higher. ISJTAs do not extend into PCAs. ATC keeps IFR traffic out of ISJTAs while in use (normally in daylight hours on Monday through Friday). VFR pilots may blithely drift right through.

Warning Area: operations conducted over international waters by military aircraft. Warning areas can be as risky as restricted areas, but, since the United States doesn't regulate the airspace, no restrictions to flight can be imposed, and you're free to fly as you wish.

Controlled-Firing Area: if you want to shoot off a rocket or blow up a mountain: contact the FAA and get an area established for the duration of your activity. Aircraft have priority over the rocketeers or demolition experts, however: they may not know about the blast. Weather has to be good: ceiling at least 1,000 feet above the highest ordinate of fire and visibility at least five miles in all directions.

Temporary Flight Restrictions: formerly called disaster areas, these are established to eliminate the air shows that used to play over a spot where a tragedy occurred. These and controlled-firing areas are the only kinds of airspace with identifiable dimensions that are not charted. (Although TRSAs are not depicted on aeronautical charts, they are diagrammed in the *AIM*.)

AIR-DEFENSE AIRSPACE

Air-defense airspace is airspace protected by the Defense Department. Potential aggressor aircraft are intercepted and given the once-over by armed fighters. There are three domestic air-defense identification zones (ADIZ): in Alaska, along the United States-Mexican border, and in the Panama Canal Zone. Coastal ADIZs are along the Atlantic, Gulf, and Pacific shorelines and at Hawaii and Guam. The distant early-warning identification zone (DEWIZ) is in the Arctic, where unreported aircraft coming over the pole are scrutinized with great interest. A flight plan is required for penetrating or operating within the DEWIZ or an ADIZ. All the air space of the United State exclusive of ADIZs is designated the *defense area*.

The airspace structure can be as simple or as complicated as you like. Low-altitude, low-speed pilots who ignore its arcane reaches can get by nicely if they maintain control-zone weather conditions wherever they fly. Lots of pilots wouldn't want to fly in poorer weather anyway. Other pilots should be familiar with every kind of airspace that their engines can breathe. Man's territorial imperative, which the post-Freudians believe to be a stronger impulse than the sexual one, is certainly borne out by what has been done to the airspace in the name of safety and progress. Chain-link fences, stone walls, and guarded borders seem natural, and when the FAA began carving up the air, they succeeded in keeping the airplane within the familiar context of the wheelbarrow and the locomotive. No one is going to be able to accuse them of thinking that the airplane is a revolutionary gadget.

40.

IFR VS. VFR

Many pilots think that IFR flying is hard and that VFR flying is easy. Others say that IFR is safe and VFR is dangerous, but a recent statistical study added a new twist − it showed that instrument-rated pilots have a higher involvement in many types of accidents, including weather accidents, than non-instrument-rated pilots. (The instrument-rated pilots were not necessarily operating on an IFR flight plan, however.) Statistical studies are easily manipulated, and one could probably easily develop a study that would produce the opposite conclusions. For practical reasons the hard-versus-easy question and the relative safety of each mode are most logically addressed by examining the methods of IFR and VFR operation. IFR is inherently easier than VFR and offers greater potential safety. The word "potential" is important because, if IFR pilots do not maintain proficiency and if they succumb (literally) to the urge to go below minimums on approaches, then IFR can be lethal, just as VFR is unhealthy for the pilot who presses on into inclement weather.

Forget the idea that every IFR flight involves parting the fog on the runway with the prop spinner; climbing through 10,000 feet of cloud; flying level through thunderstorms, hail and scattered pitchforks; and shooting a 2,400-foot runway visual range approach. Many wise pilots make it a habit to use IFR on every flight, regardless of weather, and in so doing they see the constant value of it. If an IFR-flight plan enables you to continue on-course through a small area of low visibility or through a little cloud bank or to depart on time even though the thin overcast hasn't burned off, it is the best way to go − no doubt about it. The procedural requirements of IFR sometimes take an extra few minutes, and the airways system adds a few extra miles, but if these delays are balanced against having to fly at unfavorable altitudes because that's where the VFR sky happens to be and circling and scratching while you wonder how to get around that patch of crud ahead, the IFR bird is the speediest of all. The safest too.

Planning an IFR flight is simpler than a VFR. The airways are on the Jeppesen and NOS charts, with distances and bearings charted. The minimum safe altitudes are right on the charts, and, if a special departure procedure is required to clear obstructions, it is spelled out. VFR flight-planning with sectional charts involves rulers, lines, course-measuring, and checking elevations and obstructions on the chart − to say nothing of the job of folding and unfolding and refolding that cumbersome wad of chart paper. If IFR procedures are followed, terrain clearance is automatic. The VFR pilot is on his own.

Checking the weather for an IFR trip is certainly no more demanding than checking for VFR, and, if the FAA ever pays a little attention to general aviation IFR requirements, the weather part will become simpler still. If you are flying VFR, you are concerned about ceiling and visibility over every inch of your route; IFR, your ceiling and visibility requirements are minimal, and your attention shifts to a cross-section of the air between here and there at a selected cruising altitude.

The IFR pilot tends to do a more thorough job of organizing and preparing for his flight as he settles into the airplane, which makes his flight easier and safer. Radios are set in advance, charts are properly folded, and the approach plate for the departure airport is at hand, with the airport diagram available for taxi assistance. The good instrument pilot is highly organized when he calls for his clearance, while many a VFR pilot neglects to put his brain in gear until he engages the aircraft starter or keys the mike.

The IFR man is also way ahead during the climb, too. Except on days when there are no clouds below 10,000 feet, VFR climbs for the average crosscountry either have to stop at low altitude in rough air or they have to be based on clearing clouds on the way up. This involves zigzagging and wondering if an IFR airplane is going to punch out of that piece of cauliflower ahead. The IFR pilot climbs as he is cleared, and it is usually an on-course climb to cruising altitude.

Virtually all en-route IFR flights run right down the airway, filed with little ado. All the pilot must do is change frequencies when he's told to, set the VORS, and talk and squawk as necessary. There could hardly be a simpler way to navigate and cover distance. IFR pilots seldom "get lost," and when they do start to stray from the prescribed path, the controller usually speaks up about it before the airplane is more than three or four miles from the airway centerline. On a small percentage of flights, the IFR pilot does have to evaluate and deal with what might be called marginal or hazardous en-route conditions — thunderstorms or ice problems. The IFR man still has a lot of help, though: he does not have to worry about staying clear of clouds, and he is on a radio frequency with other people who are flying in the same conditions; it is therefore often possible to learn of conditions well in advance, and, if the excitement is provided by thunderstorms, the controller sees at least some of the weather on his radarscope.

Many VFR flights are as pleasant and uneventful as IFR flights, but if the weather is marginal, VFR becomes a hazardous and lonely business. "Continued VFR flight into adverse weather conditions" is one of the leading general aviation accident causes. It's easy to see why, for when a VFR pilot reaches that point in a flight where the venetian blinds are slowly closing all around, all his options and information sources often tend to vanish at once. The clouds form a legal barrier and a distinct hazard to a noninstrument pilot; the terrain below usually presents an unfriendly surface on which to land and think things over; and the low altitude (unless he's caught on top) precludes calling anyone to get the latest scoop on the weather. The pilot is left with nothing but sweaty palms, a bad taste in his mouth, a map folded to the wrong panel, and a collection of aluminum hurling through the air uncomfortably close to the cold, hard ground. Anyone who has flown a lot of VFR cross-country knows the feeling, and the ways used to get out of such situations don't bear repeating. One or two such encounters make you double your VFR minimums — and make IFR at least twice as attractive. They also make a mockery of the idea that VFR is safer than IFR under marginal conditions.

Almost all the United States now has radar coverage, which means that the IFR aviator usually gets radar-traffic advisories all the way on every flight, controller workload permitting. This is not to say that the traffic advisories should be considered as a substitute for active eyeballs, but they do handsomely supplement your efforts to ascertain other traffic: it's a safe bet that the controller will see as much traffic as you do, even if you are a good airplane-spotter. (This service is available to VFR pilots too, but only if it doesn't interfere with the controller's handling of IFR traffic.) On arrival the IFR system is at its absolute best if the destination is a high-density airport. Show up VFR at one of those places and you have to work to get a word in on the frequency, and then you have to receive and understand a set of instructions about how to fly to the airport. With IFR you get a handoff from the center to the terminal radar controller plus the frequencies to use, and it is a very smooth procedure. IFR arrival is also easier at out-of-the-way places, too. If flight conditions are the least bit fuzzy, a navaid procedure guides you to the airport, and a set of instructions tells you what to do if you don't see that airport at a specified time. It sure beats the VFR method of flying down the highway until you pass the drive-in movie and then taking the second railroad track on the right to the airport. The navaids can be used for VFR, to be sure, but, without instrument charts that show precise bearings and minimum safe altitudes, VFR use of navaids is an approximate affair.

VFR flying still has a great future: this discussion is intended only to puncture the unpleasant myth that VFR is easy and IFR is difficult or that IFR is not as safe as VFR. The truth is that, once a pilot is properly trained, IFR is the easier of the two most of the time, and its safety potential is far greater than VFR except on clear days when they are probably equal. There will always be some situations in which VFR offers an advantage over IFR: if there's ice in the clouds and you can fly beneath the clouds VFR, that might be the best bet. If there are a lot of thunderstorms and the controller can't help with his radar or with new routing because of an IFR traffic crush, and the weather outside the storm area is good VFR, it's often best to cancel the IFR and go on VFR. A strong headwind above 3,000 or 4,000 feet can often make it more efficient to fly VFR at 1,000 feet. But most of the time IFR offers the most in safety and efficiency from start to finish — if the pilot is up to standard.

41.

AIR-FILING IFR

The horses mechanically rear, their colors blurring into the brass glints of the merry-go-round, and you pause before you grasp the painted chariot as it whirls by, thinking of how the dust must smell should you move too slowly. Filing IFR in the air is like trying to leap on a merry-go-round that's already spinning — precarious. Most air-files come from pilots who began on VFR but who encounter deteriorating weather. The first forecasts were wrong, the pilot hardly has an accurate picture of the weather in his mind, and he'd be smart to land at the nearest airport, check the weather, and get on the merry-go-round while it's stopped. But the streets of hell are paved with good intentions, and we all occasionally find ourselves peering down at the ground, trying to maintain VFR while filing an IFR flight plan. The charts are probably strewn among the four corners of the cockpit, while an open microphone records a long "Ahhh," clearly transmitting your confusion.

Air-filing is like so many other sins: you can get away with it occasionally if the commission of the sin is premeditated enough to stimulate confidence. The important thing is to decide to air file as early as possible and then to prepare your speech. Don't start calling until you're ready to give the full flight plan to the man on the ground.

Whom to call? If you need an en-route clearance, the center can sometimes handle the whole thing for you: they will take the flight plan and give the clearance. It's a good idea to experiment with them. Find the appropriate frequency for the center on the chart. The first call is a simple request — all you want to know is whether or not the center will whip up a clearance for you — so be simple: "Center, this 4215R, 30 southeast of Waterloo, will you take an IFR flight plan?" If they can handle your request, they will say so and will ask for the information they want. They usually ask for an abbreviated flight plan: aircraft type, true airspeed, fuel on board, destination, route, altitude, and time en-route. They may need only the destination, route, and altitude before issuing the clearance and request the other items later. Don't read off a whole flight plan unless they ask for it.

If the center controller does not have the time or inclination to take your flight plan, he'll refer you to a flight service station. The FSS will want the full flight plan, so have it ready when you call. A few minutes after you read it to the FSS, they will give you a center frequency to tune — or they will issue the clearance themselves and give you a frequency on which to contact the center. If you want a clearance into an airport that is served by a terminal radar facility, it

195

will almost always handle your request directly. If there's any traffic, however, you may be last on the list for service.

Another item of importance is your position. Most people who issue clearances have a radarscope in front of them, and, if they can find you on that scope quickly, they'll have a much higher opinion of your airmanship. The service will be better too. When they ask for your position, don't answer with a vague "Unnn, I'm north of the Puddephatt Intersection, about 30 or 40 miles." Identify your position, especially with center controllers, in relation to VOR stations, airway intersections, or airports.

One final item: some air-files come from pilots who dash to the airport and leave hurriedly. The IFR isn't filed before takeoff because the centers want flight plans to be filed 30 minutes before takeoff. They seem to run a de-facto enforcement campaign by waiting to issue clearances until 30 minutes after they are filed. The 30 minutes allow the computer to process the flight plan, make the flight-progress strips, and deliver everything to the right place. Even if you are going to run to the airplane and fly away as soon as you hang up the receiver, it is still better to file the IFR plan *before* takeoff. Just tell the man that you'll be leaving VFR and will contact the center for a clearance after you've taken off. If he's a nice guy, he'll call your plan in to the center. Just be sure that you can maintain VFR and that you will be in range of a center communications site before the IFR clearance becomes necessary. Better yet, board the merry-go-round with stately gait and every hair in place: file the flight plan well before you get off the ground.

42.

IFR AT NIGHT

IFR flying is procedural flying. Match the numbers on the instrument panel with the numbers on the chart, and you get where you want to go. The fact that IFR is seemingly such a cut-and-dried operation should mean that whether it's day or night has little bearing on the flying. Not true. Certain phases of *night* IFR operation involve demands greater than and different from their daytime counterparts. If night IFR is carefully studied, you will find that the tough situations and problems can result when the routine of matching numbers on the panel to numbers on the chart ceases. The same things lead to IFR accidents at night as by day, but they do so in an absolutely startling proportion in some areas. In one year's weather-related accidents five out of six mishaps involving a descent below the minimum descent altitude occurred at night, and four out of five thunderstorm-related accidents happened at night. The other problems − engines, systems, and ice − occurred in numbers reasonably proportionate to the amount of IFR flying done in the daytime and in the dark. There are two areas of unusual demand in night instrument flying on which to concentrate initially: the nonprecision approach and thunderstorms.

Another weather problem − ice − would not seem to be any more difficult to deal with at night. If anything could cause problems there, it would probably be lower surface temperatures at night, but cold nights also tend to be clear nights.

Even on a rainy night with relatively good visibility it is difficult to make accurate height judgments simply by looking outside. The most difficult situation might arise when flying in precipitation over a dark area with the lighted airport in the distance. Looking through precipitation to lights several miles ahead gives the illusion of greater than actual height, and even without precipitation a pilot would be hard pressed to peer at lights a few miles ahead and keep a running idea of the airplane's altitude.

The circling approach, usually something to be avoided, can actually be turned into a good thing on a night IFR approach with weather well above circling minimums. Rather than flying a long straight-in over an unlighted area it might be helpful to fly up over the airport at the minimum descent altitude for a circling approach and to circle the field for a landing. Why? To bring the maneuvering up over the lighted area and to allow use of a planned and familiar traffic pattern. It is more difficult to make an accurate estimate of distance to the runway from a point on the long final than it is to fly your usual downwind, base, and final.

Another night-approach aid is a good understanding of the relationship between yourself, your airplane, and the approach in use. Problems start when the pilot severs his ties with the procedures printed on the plate and starts doing things on his own. The rules specify only that you are not to leave the MDA unless the aircraft is in a position from which a normal approach to the runway of intended landing can be made and unless the approach threshold of that runway, approach lights, or other markings identifiable as the approach end of that runway are clearly visible to the pilot. That rule gives the pilot reasonably good guidelines to follow in the daytime − if you can see the runway ahead, you are unlikely to fly into something before reaching it − but it leaves the night-IFR flier in a black hole with no guidance: "You see the runway, you start down." If you are lucky and smart, you don't hit the ground until you reach that runway.

An interesting exercise to illustrate the potential inaccuracies of eyeballing it to the runway can be found on any ILS, day or night, by hood-flying. (Try it at night to get the best illustration of the phenomenon.) Shoot the approach, raise the hood at minimums, and continue visually for your landing. Ten seconds after lifting the hood and transitioning to visual glance back at the glideslope needle. Chances are that you'll be below the glideslope. Try this with higher minimums than the ILS. Raise the hood at 500 feet − a normal MDA for a nonprecision approach-transition to visual, and look back in a moment to see how your approach path compares with the electronic glidepath. You'll probably be below it, unless some extra effort was made to maintain a shallower than normal approach slope. The pilot must thus pencil some nighttime notes of his own into the margin next to the rule about not leaving the MDA until the lights of the runway are in sight and the airplane's in position for a normal approach. The pilot must formulate a plan for leaving the MDA − a plan to insure that the security of a safe altitude is not abandoned until you intercept an imaginary path that leads to the end of the runway.

Approach-path control is possible from within if a pilot engages his brain before starting down. For the sake of illustration assume a 90-knot speed for the final part of your approach. That's 1.5 nm per minute. The average light airplane is capable of descending 1,000 feet per minute, which at 90 knots would be descent rate of 667 feet per mile. If the MDA is 500 feet, that means that you do not have to leave that safe altitude until you are a mile from the end of the runway. The difficulty in judging distances at night would increase at least as the square of the distance from the runway, and, since a mile is pretty close, a pilot would be flying at maximum advantage by doing it this way. The old "fixed-spot-in-windshield" method is the best guide once the descent from the MDA is started a mile from the end of the runway. While you are flying a constant airspeed and maintaining a constant attitude, the spot that remains in a constant position in the windshield is the point toward which the airplane is descending. The approach end of the runway should thus remain fixed. If it moves lower in the windshield, you are too high; if it moves higher in the windshield, you are too low. The later is hazardous to health. Also, if the rate of descent noted as you establish an approach path that leads toward the end of the runway is less than 500 feet per minute, the approach is a shallow one. The steeper the approach, the more likely you are to clear obstructions comfortably.

Practice makes proficient, if not perfect, on the visual part of night IFR approaches, and an evening of practice will pay dividends. There is a visual distraction that plagues the pilot on both day and night IFR approaches: the ground or lights on the ground begin to appear in your peripheral vision as they slide beneath the airplane; and nothing can be seen ahead. The clear message from the horned little man sitting on your left shoulder is to descend a bit more in such a case to get contact. The guy with the halo is in the altimeter, though, and his message is the one that saves the day (or night): do not leave the safe altitude until the airport is in sight and you are close enough to ensure terrain clearance on a descent from present position to the end of the runway.

One piece of hardware available to the IFR pilot who wants a little extra help is the radar altimeter, which will give height above the ground and can also be used as an alert system. It won't tell you anything the approach plate, the regular altimeter, and your brain won't tell you, but it can help by giving a true, absolute, and instant indication of the space between the bottom of the airplane and the top of the ground.

Beware of the traps of weather in night approaches. If you have done everything right, are continuing down toward the runway on a good slope, and the runway disappears, go to the missed-approach procedure. It's poor practice to sit there in a descent below the MDA after a piece of cloud gets between you and the runway. A special time to be alert for this possibility is when any scattered clouds are reported below the ceiling: 200 scattered, 800 overcast might be one of the toughest possible weather situations for a night approach, for that 200 scattered could be in just the wrong place. Also be aware that weather observations taken at night are generally not as accurate as those taken in the daytime.

Airplanes often hit the ground on final, and they often reach the ground *before* even coming to the point at which a descent should be started to reach the runway and before reaching a point at which the runway could be seen under the existing weather conditions. Leaving the MDA prematurely is more lethal at night than in the daytime simply because the chances of belting the ground without getting any visual clue of impending disaster are obviously much greater in the dark. Another positive step might be to raise your personal minimums at night *unless* a lot of night IFR flying is done. Success at nonprecision night approaches probably depends more on an understanding of the particular task and experience at it than anything else, so a pilot's judgment of his night-approach capability should be based more on his knowledge and experience in that specific area than on his total experience. In other words, if you've never shot one, work up to night approaches to minimums very gradually.

Thunderstorms are a top IFR problem at night, though this may at first seem contrary to logic. Can't you see the lightning from the storms at night and just avoid those areas? Apparently not, for the record clearly indicates that IFR pilots without airborne weather radar have more trouble with thunderstorms at night than during the day. Perhaps the assumption that lightning will define the impossible areas leads pilots into the arms of the devil himself. If you think about the way in which the radarless pilot flies (successfully) in thunderstorm areas by day, the hazards of the cumulonimbus at night become apparent. In the daytime you use a combination of information from the controller on the ground and what you can see out front. If you had to make a trip through an area of storms, would you rather do without your vision or without the controller's inputs?

The choice is made for you at night, and the constant limitations of the ground-based radar couple with night visual limitations to make flying in thunderstorm areas a more difficult thing to do. Of course the lightning illuminates the storm and you might be able to stay away from that part of it by referring to what you see when the lightning pops, but what about that building cell that is just about to mature. There's no lightning until one does mature, but here is plenty of turbulence to work you over thoroughly. Once you get into the midst of an area of thunderstorms, the lightning turns out to be more confusing than helpful. The IFR pilot needs to be aware of the difficulties of navigating through an area of thunderstorms at night and should make adjustments to compensate for the lack of good visual cues to use in conjunction with reports from the radar controller. When in doubt, let lightning help: make a standard-rate turn until the lightning is behind you, then fly straight ahead (with ATC's blessing, of course).

Mechanical problems can be a bit more serious for the night IFR pilot, but it is hard to find any measurable level of exposure to hazard in this area in the accident reports. Perhaps the people who fly night IFR match their faith in the machine with dollars at the shop. As a matter of good practice the IFR pilot flying at night should pay particular attention to monitoring his systems as he drones through the sky. In a single, an alternator or generator failure, for example, is likely to be twice as serious at night — at least it would leave half the time to do something because of the additional drain of the lights. You'd want to catch it as soon as it failed. In a twin, you'd also want to learn quickly of any failure of half a redundant system.

The night IFR single-engine guy might practice some preventive maintenance on systems. A new battery every two years is not a bad idea to ensure that a good battery will be at hand to last through an approach and landing if the alternator or generator should fail. It has also been the authors' experience that alternators are good for an average of 1,000 hours and a maximum of 1,500 hours, and exchanges on these are inexpensive enough to make replacing old ones a reasonable proposition. Vacuum pumps have a life expectancy of 500 to 1,000 hours. (None of these are bad ideas for day IFR pilots either.)

The mechanical condition of the pilot is important too, and here there are some difficult questions. Night en-route IFR should not place any more load on the pilot than day VFR flying if the weather is horizontal (as opposed to vertical, as in the case of thunderstorms). At least there is no sign in the record that pilots have en-route IFR problems peculiar to nighttime other than in thunderstorm areas. There is a condition to this, though, more related to the time of day than to the condition of the light: unless a pilot is a true night person, he's conducting his flight on the backside of his personal energy curve. The night IFR pilot must be aware of that. He should use oxygen on any reasonably long night flight if the cruising altitude is above 5,000 feet, and if there's not enough in the bottle for the whole flight, he should file for an altitude as low as practical and sharpen up with the oxygen for about 30 minutes before reaching the destination.

The approach comes at the end of the flight when one's level of fatigue is peaking. There's no way to turn this around and shoot the approach before takeoff, so the pilot must be aware of the task and make an extra effort to fly methodically and to check every number and setting carefully. The strong urge to complete the approach, get to the car waiting obediently at the airport, and drive home must be dismissed. The night approach must be an impersonal thing. If everything isn't just right and if the runway is not visible when it should be, the verdict has to be totally in favor of a trip to the alternate. Another night in a motel might be a pain in the tail, but the old oak tree short of the runway could be far more painful.

There is one final area that must be discussed, even though it does not strictly relate to instrument flight. It is the role of the instrument-rated pilot in VFR weather accidents. Surprisingly, this is quite a problem. Almost twice as many instrument-rated pilots are involved in night weather accidents when flying VFR than are involved when flying IFR. If you examine the record, it is apparent that the instrument-rated pilot who flies VFR in marginal weather at night does little better than the nonrated pilot. The rating is of little value unless a pilot is on an instrument flight plan and flying a clearance, and most rated pilots involved in night VFR weather-related accidents would have survived if only they had filed.

Night IFR can be useful, and it is a perfectly sensible way to travel if a few differences are recognized and respected. The record strongly suggests that the pilot flying IFR at night must be wary of going below minimum descent altitude on approach and that a little extra be added to the minimums on nonprecision night approaches. The record also indicates relatively more difficulty in dealing with thunderstorm activity at night in airplanes without airborne radar. If a pilot minds his manners in those areas, maintains his airplane properly, doesn't overestimate his personal stamina, and devotes a little time to practicing night flying, there's no reason why a night IFR flight shouldn't be as easy and pleasant as a day IFR flight.

43.

THE MISSED APPROACH

The missed approach is by far the most difficult maneuver that an IFR pilot is ever called upon to perform. It always comes as the unhoped-for climax of a tight approach, and it requires a total and instantaneous reversal of the pilot's objectives just as he is at the most critical altitude and in the most critical configuration. At one moment he's descending quietly, taking quick looks out the windshield for a runway that he knows is there and fully expects to see with each glance; in the next breath he's surrounded by the roar of a full-throttle climb, an out-of-trim airplane, and the guilty sensations of one who has just tried and failed.

The consequences of not performing the maneuver properly are recorded in accident statistics, and surprisingly often the guilty pilot is an ATP professional and with a right-seat backup. This means that cockpit discipline and mental attitude are more often the weak links than are skill and technique, and the most unnecessary missed-approach accidents are those that result from a breakdown in discipline. After an air-carrier accident a year ago in which the aircraft flew into some beach cottages off the approach end of the runway, cockpit tapes were played back; passing through the decision height the copilot was heard to call out to the captain that he was still "sinking five" with the airport not in sight. In a moment the copilot said, "This is too low," to which the captain replied, "I can see the water − I got straight down." Copilot: "Ah, yeah, I can see the water − man, we ain't 20 feet off the water." The next sound on the tape was of crumpling metal. Last year another crew hit a hill three miles short of the runway. Another hit wires on the airport boundary. Another hit the approach lights. These are usually called approach accidents, but they are actually missed-approach accidents. In each instance the pilot was in a missed-approach situation, but at his decision height discipline went out the window and he failed to switch his thinking from get-down to get-up, which is the most critical and most often committed mistake in a missed approach.

Maintain strict cockpit discipline, and that kind of accident will never happen to you. All you have to do is resolve to never go below minimums unless the runway is in sight and the aircraft is in a position to land on it. This can be a difficult resolution to follow, because a hole may open up directly below through which you may see a familiar landmark or perhaps a piece of the airport itself, and you'll have an overpowering urge to duck under and complete the approach visually. If you can actually see the airport, instinct will make you want to close the throttle, put on the full flaps, and dive at the runway. Don't. The insidious thing about the duck-under or dive-for-the-runway approach is that you may get away with it for years before you dive under one day and discover that it's only the hole that goes all the way to the ground. Or you may discover that you can see down through a thin fog but can't see through it horizontally. That is what happens to those trained professionals: their years of experience are terminated by one last discovery that what works most of the time doesn't work every time.

Pilots who fly little airplanes have a particular cockpit-discipline problem. Passengers, in their urgent wish to be helpful, can sometimes talk them into a bad situation. On a tight ADF approach one night a lady in the back seat suddenly shouted, "Here's the runway back here." The author instinctively looked back, and, sure enough, the runway end lights were just behind the left wing. A missed approach begun on a foggy night with you looking back over your left shoulder has no joy in it, so, before beginning a tight approach with passengers, tell them to keep their mouths shut until they hear the squeak of tires on the runway. If you have a pilot who you can trust in the right seat, you might ask him to watch for the approach lights and the runway through the windshield, but he shouldn't utter a peep until they are firmly in sight. By no means should he or you pay any attention to what might be seen out the side windows. Until you can see the runway through the windshield over the cowling while you are in a normal glide, you can't possibly land on it.

An important corollary to cockpit discipline is to choose your own personal minimums below which you will not wander. The minimums printed on approach plates are not for every pilot. They presume a certain skill at executing a missed approach, and a certain level of imperturbability, that the new IFR pilot may not have. He needs a little extra cushion. Even airline captains moving into a new type of aircraft or to the left seat after years on the right side have higher minimums for the first 50 to 100 hours.

Many new instrument pilots choose arbitrary numbers − such as 500 and one, or 200 feet and a half mile above published minimums − for their personal minimums. That can lead into a trap. Terrain and proximity to antennas or buildings can make 500 and one, or plus 200, too thin a cushion. A better plan is to use circling minimums for the first few dozen actual approaches. These do allow for obstructions in the vicinity of the airport, so the missed approach must be badly mangled before serious trouble is encountered. (On some runways, however, circling minimums are actually lower than straight in. In that case add 20% to the straight-in minimums.) The new pilot should use circling minimums even on a full ILS approach − *especially* on a full ILS approach, in fact. The reason is that more altitude will probably be lost in a missed approach in the transition from a glideslope descent to a climb than from a nonprecision descent. Analyzing the pilot's workload on the two types of approaches will reveal why this is so. (The novice should be wary of shooting glideslope approaches at all during his first few approaches in actual conditions. The chances of making a serious blunder are much less in descending to an indicated altitude than in chasing a glideslope needle.)

Having the proper mental attitude goes hand-in-glove with cockpit discipline. It's safe to say that every duck-under accident is the result of a pilot who is more willing to accept the consequences of busting minimums than the consequences of a missed approach. This might be simply the pressure of pride. If it is, it's the worst kind of false pride. Many veteran IFR pilots will tell you unashamedly that, as time goes on, they make more and more missed approaches. The pressures can be even more compelling than pride: low fuel, fear that the alternate will be as bad as the airport being approached, a load of ice, lack of preparation for a missed approach. Developing the proper mental attitude is as simple as beginning every IFR flight with the expectation that the destination approach will be a missed one. Have plenty of reserve fuel; have not one but two alternates that you are confident will be good; don't continue letting down into ice until the airplane is carrying so much that it can't get up again; study the missed-approach procedure and have the routing, altitude, and first navigation fix to your alternate written down. If you have an extra navigation radio, set it up for the missed-approach procedure before beginning the approach itself. In short, don't let yourself get squeezed into a corner from which the only escape is a landing. That is *the* coffin corner in IFR flying.

Following this rule will take the pressure off and make the actual technique of a missed approach as simple as rice pudding.

As with every other IFR procedure, preparation for the execution of a missed approach must begin on a bright, sunny day with the visibility stretching into tomorrow. Go up to about a thousand feet (higher than that and the airplane will respond differently to go-around power), set up a standard instrument-approach configuration and rate of descent, go to missed-approach power, and begin taking careful note of precisely what happens as you transition to climb. How many swipes at the trim wheel are required? How many "counts" does it take to bring the flaps from approach to climb position? Which sequence yields the most immediate and certain rate of climb in *your* airplane with *you* flying it: flaps and then gear or gear and then flaps? (Each aircraft model is different, and each airplane may be different in the hands of different pilots.) What is the *precise* pitch attitude shown on your artificial horizon at the beginning of a missed approach? As the flaps and gear come in? Lightly loaded? With all seats full? How much roll error does your horizon have in going from the approach to climb configuration? Which direction is it? (All artificial horizons have acceleration errors, which vary with the rate and magnitude of the acceleration. It shows up as roll error in straight-ahead flight, both roll and pitch in turns.) Go through this exercise at least a dozen times, both visually and with a hood and check pilot in place until you've got it all down pat. Now you're ready for some practice actuals.

That may shake you up: practice actual missed approaches? Hardly anyone flies into an airport in which the chances are less than 50/50 of getting in, but until you've actually missed a few approaches, you're not really a complete IFR aviator. Because we tend to shy away from marginal IFR, many of us fly instruments for years before we have an actual. As a consequence, we make even the easy approaches with sweaty palms, because the possibility of a missed one hangs menacingly in the regions of the unknown. This will pass after a few actual missed approaches, but it's better to experience them at your convenience under carefully selected weather conditions than at the whim of Dame Fortune.

The perfect weather conditions for missed-approach practice are low ceilings (about 100 feet below your personal minimums but no lower than 300 feet agl), very flat terrain with good visibility underneath (2 miles or more), low tops, and excellent VFR within 50 or 100 miles.

The good visibility underneath is to give you confidence that you aren't going to fly into the ground without ever seeing it and to make the approaches legal. FAR 91.116(b) allows us to shoot all the approaches we want when the ceilings are below minimums, provided the visibility is above published minimums. On those kinds of days the traffic is light, so you probably won't be inconveniencing other pilots, particularly at the smaller airports. Although it's embarrassing to shuttle back and forth between the good weather and bad, don't hesitate to shoot a couple of approaches, then to go home for fuel and a cup of coffee. You can take an instructor with you if you like, but it's best if you do it ourself. A trusted copilot won't hurt, but, since the object is to gain confidence, you should be the undisputed pilot in command.

You'll find that actual missed approaches are relatively simple after your VFR practice. If you follow the cardinal IFR rule − aviate, navigate, communicate, in that order − you won't go far astray. To commence the go-around, first go up on the power and simultaneously plant your eyes on the artificial horizon. Level the wings, set the proper pitch angle, then, with a quick glance at the turn-and-bank, center the ball. Failure to follow this sequence is the most common and most hazardous mistake of low-time IFR pilots. They are over-concerned about airspeed, altitude, and heading at this critical moment, and their eyes dart from the altimeter to the DG to the rate of climb to the airspeed as they pump the wheel and try to make everything move in the desired direction. If it does nothing else, the VFR practice should convince you that, if you go to the artificial horizon first, level the wings, and establish the proper pitch angle, everything else will fall into place without having to look at the other instruments. You can then devote your attention to getting the airplane cleaned up, the trim set, and the cowl flaps open.

Getting the wings level is first on the list of things to do with your eyes. The reason is that the most critical missed approaches usually begin with the pilot looking for (or at) the runway. The airplane is also often placed in a bank in a futile attempt to get lined up on a fuzzily seen runway. You want to give yourself something simple and useful to hang onto for an instant while you rearrange your thoughts for the climb. Leveling the wings will do you the most good at this instant. This stops any turn that may have developed as you were looking outside the cockpit or making one last stab at the localizer and puts the airplane in the best attitude for climbing. As the wings come level, you must without delay set a pitch attitude that experience has taught you will result in immediate climb, then glance at the turn-and-bank indicator. The ball should be centered, because airplanes climb best when the controls are coordinated, but also check the turn needle to be certain that it agrees with the artificial horizon. This is no time for a delay in discovering that your vacuum system has failed. If it is a partial panel go-around, incidentally (which you have previously practiced, of course), knowledge of your airplane and precisely how to trim it for a smooth and certain transition to climb will make all the difference. Chasing lagging airspeed and rate of climb on an IFR go-around can be exciting. After the airplane is leveled and a positive climb is established, begin the navigating phase. Turn to the first heading called for in the missed-approach procedure, and, if you didn't set your VOR or ADF for a missed approach earlier, do it now. The primary thing to remember, though, is to aviate − in other words, climb. Navigation is secondary to that.

Communication is last on the list of things to do. Don't wait until you've gone to the fix and established a holding pattern before you tell ATC that you've missed the approach but do give yourself a minute to get the airplane firmly under control and the new navigation situation firmly in mind. Then decide what you're going to do next and, finally, pick up the mike.

If you saw a piece of the airport but weren't in a position to land on it or if the missed approach was due to straying from the final approach course, you may elect to shoot another approach. You needn't be embarrassed about this. Grizzled airline captains sometimes shoot several before getting in or giving up. If another approach would be a waste of time and fuel, go to your alternate. In either case it's more professional to make a decision before communicating with ATC, so you can tell them on the first call what you're going to do next. Otherwise they'll come right back after you've called the missed approach and ask, "What are your intentions?" This tends to put pressure on you to make a hasty decision, and hasty decisions have no place in IFR flying.

If you break the missed approach into its three natural parts — preplanning, technique, discipline — and tackle them one at a time before the approach commences, the missed approach will never be a problem.

INDEX